SpringerBriefs in Earth Sciences

SpringerBriefs in Earth Sciences present concise summaries of cutting-edge research and practical applications in all research areas across earth sciences. It publishes peer-reviewed monographs under the editorial supervision of an international advisory board with the aim to publish 8 to 12 weeks after acceptance. Featuring compact volumes of 50 to 125 pages (approx. 20,000–70,000 words), the series covers a range of content from professional to academic such as:

- timely reports of state-of-the art analytical techniques
- bridges between new research results
- snapshots of hot and/or emerging topics
- literature reviews
- in-depth case studies

Briefs will be published as part of Springer's eBook collection, with millions of users worldwide. In addition, Briefs will be available for individual print and electronic purchase. Briefs are characterized by fast, global electronic dissemination, standard publishing contracts, easy-to-use manuscript preparation and formatting guidelines, and expedited production schedules.

Both solicited and unsolicited manuscripts are considered for publication in this series.

Marzhan Nurpeissova • Meiram Begentayev
Yerzhan Kuldeyev • Ruslan Nurlybayev

Development of Technology for Processing of MMC Waste

Creating New Building Materials

Springer

Marzhan Nurpeissova
Satbayev University
Almaty, Kazakhstan

Meiram Begentayev
Satbayev University
Almaty, Kazakhstan

Yerzhan Kuldeyev
Satbayev University
Almaty, Kazakhstan

Ruslan Nurlybayev
Satbayev University
Almaty, Kazakhstan

ISSN 2191-5369 ISSN 2191-5377 (electronic)
SpringerBriefs in Earth Sciences
ISBN 978-3-032-12975-8 ISBN 978-3-032-12976-5 (eBook)
https://doi.org/10.1007/978-3-032-12976-5

This Springer imprint is published by the registered company Springer Nature Switzerland AG
The registered company address is: Gewerbestrasse 11, 6330 Cham, Switzerland

Preface

The Republic of Kazakhstan is one of the largest republics of Central Asia, with an area of more than 2.7 million square kilometers, a population of more than 20 million people, and land borders with Russia, China, Uzbekistan, and Kyrgyzstan. The bowels of Kazakhstan store a significant number of types of minerals. On the basis of explored reserves, a powerful mining industry has been created for the extraction and processing of ores of ferrous, non-ferrous, and noble metals of various types. The mining and metallurgical complex includes more than 200 mining and processing enterprises, the sale of marketable products of which is currently carried out in European countries, the USA, China, South Korea, Singapore, Malaysia, and other countries.

The possibility of sustainable development for Kazakhstan is largely determined by the level and prospects for development of the mineral resource complex of the republic. The political, economic, and social position of Kazakhstan in the world community in the twenty-first century will depend on how it will be possible to increase and effectively manage the wealth contained in the subsoil.

A key factor for success in the competitive environment of Kazakhstan's fuel and energy sector, ferrous and non-ferrous metallurgy, and gold mining industry—the backbone of the national economy—remains the strength of its mineral resource base. Leveraging the country's explored reserves of various mineral raw materials, a robust mineral resource complex has been established, encompassing enterprises in ferrous and non-ferrous metallurgy, as well as coal and chemical industries. These enterprises produce more than 60 types of products, including pure metals, alloys, oxides, concentrates, and rolled ferrous and non-ferrous metals of various profiles, as well as gold and silver.

For the first time in Kazakhstan, extended analysis and forecast of development of the republic's mineral resource complex until 2030 is being conducted, with a compilation of geological and economic models for the most important types of minerals.

Despite the strict demands of the global market, Kazakhstan continues to play a crucial role in the world's mineral and raw material balance. The country ranks among the top ten nations in terms of reserves of lead, copper, oil, chromium, manganese, tin, gold, phosphorites, boron, and potassium salts.

Kazakhstan is a vast repository of minerals, containing all chemical elements of the periodic table. In terms of ore reserves, it is comparable only to Australia and South Africa. The country holds a leading global position in the availability of key industrial metals, including iron, lead, copper, zinc, gold, uranium, manganese, chromium, and nickel.

Since gaining independence, Kazakhstan has developed powerful companies specializing in the extraction and processing of mineral ores. By leveraging modern technologies, these enterprises focus not only on domestic projects and research organizations but also collaborate with international partners to integrate advanced technological solutions.

Mining is a vital sector of material production, directly influencing economic strength and independence of nations. It supplies essential natural resources that serve as a foundation for the manufacturing industry, enabling production of both industrial equipment and consumer goods.

Beyond resource extraction, mining projects create employment opportunities, drive infrastructure development, and contribute to improving living standards.

High rates of extraction and processing of mineral raw materials have led to the formation of a large amount of mining waste. At the enterprises of the mining industry of Kazakhstan, there are more than 20 billion tons of technogenic mineral formations (TMF) in dumps and tailings. The most important promising direction of our research is the assessment of TMF as a source of environmental pollution and as a source of secondary raw materials.

The purpose of our work is to conduct a comprehensive study of TMF in mining regions, both as an assessment of environmental pollution and as a potential source of secondary raw materials. Results of this study enable us to enhance efficiency and environmental safety of technogenic waste utilization. This approach supports industrial and innovative development of the state while promoting responsible use of natural resources and environmental conservation.

The monograph is of interest to mining engineers, researchers, master's and doctoral students in mining, construction, and environmental specialties.

Almaty, Kazakhstan

Marzhan Nurpeissova
Meiram Begentayev
Yerzhan Kuldeyev
Ruslan Nurlybayev

Introduction

Every year, in his Address, the President of Kazakhstan K. K. Tokayev draws our attention to the effective and rational use of natural resources, introduction of modern technologies and programs implementation of economic, social, and environmental processes that are implemented in life. Undoubtedly, the future of the country will be blurry if there is no balance between economy and environment. Therefore, one of the main priorities of Kazakhstan is its sustainable development.

Kazakhstan pays considerable attention to environmental issues. A proof of this is the closure by our Republic of the world-famous, large nuclear test site, stimulation at the international level of programs to preserve the Aral Sea and protect the Caspian Sea, as well as its many years of work to protect the environment from the harmful effects of technogenic disasters. By trying to improve the economic performance of today, tomorrow we will be spending ten or even a hundred times more resources to eliminate the damage caused.

The words of the President "...in accordance with International standards, environmental protection and ensuring environmental safety..." became the basis for development of environmental policy of the Republic of Kazakhstan. Considering the global financial crisis and the need to comply with environmental standards and environmental safety, the Republic of Kazakhstan is making a transition to sustainable development according to the concept for the period 2020–2030. Serious environmental problems within the framework of this program and ways to solve the problem of industrial waste are being considered.

The main vector of industrial development in Kazakhstan is the development of subsoil resources. The Republic of Kazakhstan is one of the largest republics in Central Asia in terms of the level of industrial development, which is largely determined by its powerful raw material base. The mineral resources used to produce various materials and products are widespread throughout the republic. The subsoil of Kazakhstan stores a significant number of types of minerals. Based on explored reserves, a powerful mining industry has been created for the extraction and processing of ferrous, non-ferrous, and precious metal ores of various types. The mining and metallurgical complex includes more than 200 mining and metallurgical enterprises, processing enterprises, the sale of commercial products of which is currently carried out in Europe, the USA, China, South Korea, Singapore, and Malaysia.

Mining provides countries with the necessary natural resources, on the basis of which industry creates means of labor and consumer products, provides the population with work, builds various infrastructure facilities, and also improves the standard of living of the population.

Products from mining and processing enterprises in the region (construction sand, crushed stone, brick, facing stone, marble, microcalcite, granite, serpentinite, porphyrite, clays of various compositions and other types of raw materials) are supplied not only to local construction enterprises, but also to enterprises in China and Russia, republics of Central Asia, and other countries.

Construction is one of the main sectors of the country's economy. The construction of buildings, engineering and industrial structures, and the construction of roads require various building materials. The building materials industry faces serious challenges, which consist not only and not so much in increasing the output of materials and products, but in improving their quality and expanding production of new effective materials and products that reduce material and labor intensity of construction.

The growing scale of construction in Kazakhstan requires a significant amount of mineral raw materials for building materials industry. Intensification in this direction involves the use of industrial waste instead of primary natural resources to reduce the cost of building materials. Use of solid mining waste in building materials industry is more economical compared to the production of building materials based on special mining of mineral raw materials.

This monograph is devoted to the solution of processing technogenic waste from mining and metallurgical complexes to obtain popular building materials, carried out by scientists of Satbayev University.

This monograph provides data on global environmental problems, including the state of the environment in regions of mining and metallurgical complexes (MMC) of the Republic of Kazakhstan, primarily related to management of industrial waste utilization. The main directions of waste management and technological schemes for resource-saving technologies for the production of building materials are presented.
This monograph is intended for specialists working in the mining and metallurgical industry, as well as bachelor's, master's, and PhD students in the specialties of "Mining", "Ecology", "Building materials", and "Environmental management".

ISBN 978-601
Research was carried out with the financial support of Ministry of Science and Higher Education of the Republic of Kazakhstan (Grant No. BR21882292—"Integrated development of sustainable construction industries: innovative technologies, optimization of production, effective use of resources and creation of technological park").

Acknowledgments

Research was carried out with the financial support of the Ministry of Science and Higher Education of the Republic of Kazakhstan (Grant No. BR21882292—"Integrated development of sustainable construction industries: innovative technologies, optimization of production, effective use of resources and creation of technological park".

We sincerely thank all the scientists who contributed to this monograph, sharing their knowledge and providing valuable assistance during its writing. Our gratitude extends to Zhumadilov Zhanar, Bek Ayman, and Ashimova Ainash (Satpayev University), Orynbekov Elzhan, Estemesova Aksay, and Altayeva Zaure (research associates of the International Educational Corporation), as well as Lozinsky Vasily (Dnipro Polytechnic, Ukraine). Their contributions and publications have been integral to the development of this work.

Competing Interests The authors have no competing interests to declare that are relevant to the content of this manuscript.

Contents

About the Authors

Marzhan Nurpeissova is a Professor at the Department of Mine Surveying and Geodesy, Satbayev University (Almaty). She is a Doctor of Technical Sciences and is author of over 300 scientific publications, with an h-index of 15 in Scopus and 4 in Web of Science. She holds 21 patents (KSRO and RK) and has authored 10 monographs, and 12 textbooks. Her research interests include mining, geodesy, geoecology, building materials, and industrial and environmental safety in subsoil development.Nurpeisova M.B. is a three-time recipient of the State Grant *"Best University Teacher of the Republic of Kazakhstan"* (2006, 2011, 2024), a two-time laureate of the *"Golden Hephaestus"* award in the *"Scientist-Teacher of the Year"* category (2012, 2018), and a member of the Writers' Union of Kazakhstan.

Meiram Begentayev is Chairman of the Board, Rector of Satbayev University (Almaty). He is Kazakhstani state and public figure, Doctor of Economics, Professor, Honorary Worker of Education of the Republic of Kazakhstan, and Laureate of the State Youth Prize *"Daryn"* of the Government of the Republic of Kazakhstan.He is the author of over 100 scientific publications, including 23 indexed in Scopus and Web of Science, 10 patents, and 5 monographs (h-index—5).

Yerzhan Kuldeyev is a Professor and Member of the Board, Vice-Rector for Corporate Development and Strategic Planning at Satbayev University (Almaty). He holds a Ph.D. in Engineering. He is the author of over 200 scientific papers, including 62 articles in journals indexed by Scopus (h-index—15), 11 articles in Web of Science (h-index—6), 11 patents of the Republic of Kazakhstan, 6 monographs, 7 textbooks, and 20 teaching and methodological guides.

He is a Developer of the State Educational Standard of the Republic of Kazakhstan. Kuldeyev Ye. is a Laureate of the State Scientific Scholarship for Young Talented Scientists of the Ministry of Education and Science of the Republic of Kazakhstan and has been awarded the title *Honorary Worker of Education of the Republic of Kazakhstan* (2009). His research interests include geology, geophysics, hydrogeology, water resources, metallurgy, and building materials.

Ruslan Nurlybayev is a Research Professor at Satbayev University (Almaty). He has a Ph.D. in *Nanomaterials and Nanotechnology* (specialty 6D074000). He is author of over 100 scientific works, including 12 articles in journals indexed by Scopus (h-index – 4), 11 articles in Web of Science (h-index – 5), over 43 articles in journals of the Committee for Control in Education and Science of the Republic of Kazakhstan, and 20 presentations at international conferences (UAE, Germany, Russia, etc.). He holds 11 patents and is an academician of the International Academy of Informatization of the Republic of Kazakhstan and Chairman of the Council of Young Scientists at Satbayev University. His research interests include metallurgy, nanomaterials, and building materials.

Abbreviations

AD	Automated database
ACP	Akzhal concentrating plant
Achpolimetal	Achisai polymetallic enterprise
ASM	Ash and slag materials
BASD	Bitumen-ash-slag binders
BAT	Best available technology
CP	Concentration plant
CLCTBM	Central laboratory for certification testing of building materials
DBM	Dry building mixtures
FS	Feasibility study
IUMR	Integrated use of mineral resources
IEC	Industrial environmental control
KazNITU	Kazakh National Research Technical University (Satbayev University)
KCP	Kentau concentrating plant
LOI	Loss on ignition
MPE	Maximum permissible emissions
MMC	Mining and metallurgical complex
MW	Mining waste
PMP	Physical and mechanical properties
PV	Paints and varnishes
WCC	Waste classification catalog
WD	Waste database
SMR	Solid mineral resources
SMR	Secondary mineral resources
SCMR	State Commission for Mineral Reserves
SWC	State waste cadastre
SMW	Solid municipal waste
TD	Technogenic deposit
TMF	Technogenic mineral formations
TM	Technogenic minerals

TMR	Technogenic mineral raw materials
TI	Technological indicators
ZhezCP	Zhezkazgan concentration plant
XRA	X-ray fluorescence analysis
XRF	X-ray fluorescence spectral analysis

Ecological Problems of XXI Century and Mineral Resources Exploration

Abstract The title of the first chapter sounds fashionable—now the twenty-first century is mentioned everywhere, and therefore we focused on this and called the chapter "Environmental problems of the 21st century and the development of mineral resources". Why? Because development of mineral resources and associated impact of our civilization on nature are one of the components of the overall technogenic impact. Over the past 80 years, during the life of just one generation, more than 300 billion tons of fuel have been mined and consumed in the world—as much as in all the previous years, centuries and millennia of the existence of human civilization.

Hence, on the one hand, the predicted depletion of resources within tens (at most hundreds) of years, and on the other hand, the unpredictable reaction of nature to such an active impact (waste plus disruption of the natural environment).

If we analyze the total impact of human activity, primarily the so-called developed countries, on Nature, we can conclude that the situation is approaching a catastrophe. Therefore, they begin to talk about the East-West (Russia—USA) North-South (developed and underdeveloped countries) confrontation, and about the "golden billion people" who must survive the Man-Nature crisis, in the context of rapid population growth on our planet.

Purpose of this chapter was to increase level of understanding and assist in solving global problem of ensuring environmental safety of subsoil development.

The chapter considers following aspects of the problem under consideration:

- global environmental problems associated with subsoil development;
- manifestation of environmental problems in the mining and metallurgical complex of Kazakhstan;
- on ensuring environmental safety;
- on waste management in the Republic of Kazakhstan;
- on the National Project "Green Kazakhstan".

Thus, subsoil has been and will remain in the twenty-first century the main source of national wealth for most states, and for the Republic of Kazakhstan—also a potential base for bringing the country's economy out of crisis. Therefore, one of

M. Nurpeissova et al., *Development of Technology for Processing of MMC Waste*, SpringerBriefs in Earth Sciences, https://doi.org/10.1007/978-3-032-12976-5_1

the main tasks of the mining and metallurgical industry of Kazakhstan is the development of effective measures to prevent environmental pollution through the processing and disposal of mining and metallurgical waste.

1 Global Environmental Problems and Subsoil Development

The twentieth century was marked by a sharp aggravation of contradictions in the sphere of «man—nature—resources». Without resorting to extremes, it is impossible to ignore pessimistic forecasts suggesting that relentless economic growth, amid finite natural resources, could ultimately lead to a global environmental catastrophe.

Over the past 60 years alone, over the life of just one generation, more than 250 billion tons of fuel have been produced and consumed in the world—the same amount as in all previous years, centuries, and millennia of the existence of human civilization. Hence, on the one hand, predicted depletion of resources within tens (maximum hundreds) of years, and on the other, unpredictable reaction of nature to such active energy impact (waste plus a violation of the energy stability of nature). Even 40 years ago, forecasts of the so-called Club of Rome (Professor Peccei) about environmental degradation were fulfilled and the famous book "The Limits to Growth" was published. Thus, the first warning was sounded in 1972, but little has been done since then (Fig. 1).

In the report "Limits to Growth" (Meadows 1972) and in other works (Galiev 2015; Melnikov 2003; Subsoil 2012; Enviromental 2016) draw public attention to

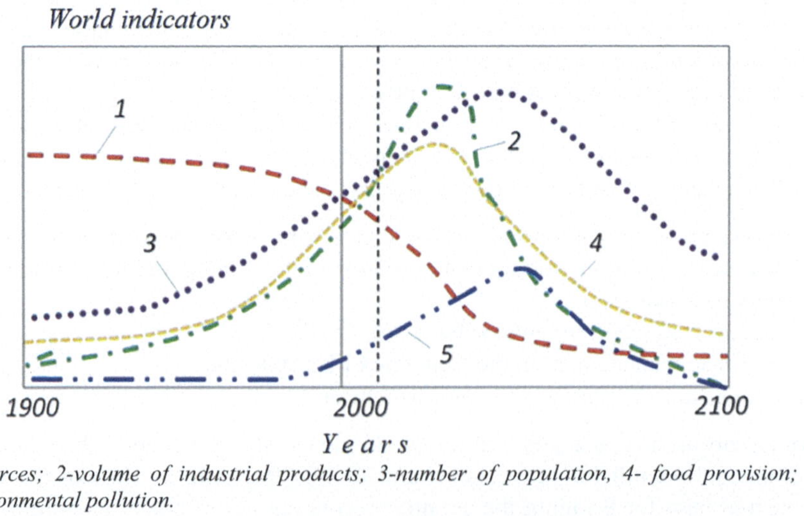

Fig. 1 Changes in indicators according to the forecast of the Club of Rome (1972). 1-resources; 2-volume of industrial products; 3-number of population, 4- food provision; 5-environmental pollution

the depletion of resources. From the latest research, let's name the publication of Professor Yoda (Japan) "The Trillema Threatening the World."

According to the definition of Japanese professor Yoda (1999), a trilemma is emerging that could lead to a global crisis within the next 50–60 years, posing a significant challenge for the Earth's population.

The trilemma (Fig. 2) consists of three interacting constraints:

- natural resources, energy and food;
- economic development.
- environment.

And it all starts with the fact that in the middle of the XXI century the world's population will reach 10 billion, according to extrapolation calculations. To satisfy the basic needs and at least minimum requirements of such population, it is necessary to have fivefold growth of the world economy. Let us emphasize again: in a little more than half a century, humanity will double, and economic activity during this period must be increased fivefold.

Economy must grow to support population growth, but its growth fivefold will entail enormous resources consumption, especially supply of food energy and water, which in turn will have irreversible impacts on the environment. Soon, with

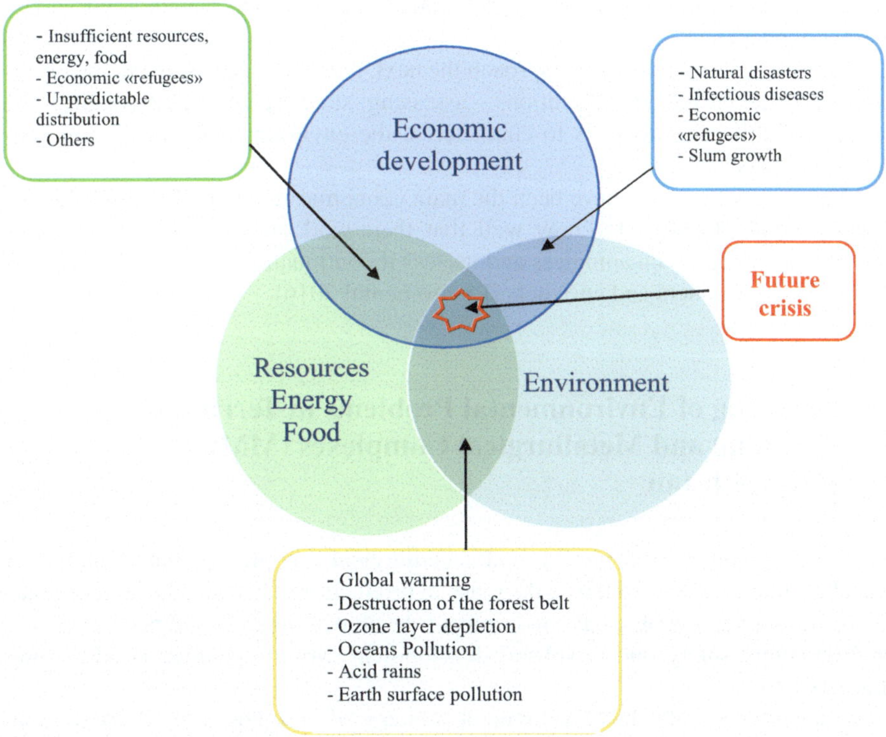

Fig. 2 Trilemma structure

a population explosion, we will be caught in a global crisis in the middle of three intersecting factors—economy, energy and resources, and environment.

Thus, environmental pollution generated by modern technology destroys health and leads to reduction in the average human life expectancy, thereby reducing rate of increase in the earth's population.

But means of reducing emissions are also purely technical. Scientific and technological progress in the field of medicine increases average life expectancy of people and contributes to accelerated population growth. The rapid growth in consumption of fuel and mineral raw materials with modern technology may ultimately become a limiter on the growth of the Earth's population. But the means to get out of this predicament are also of a purely technical nature, such as: creation of resource-saving, waste-free (low-waste) technologies for ore mining and processing, integrated use of deposits, etc.

For several previous decades, policy was mainly to increase production volumes and obtain purely economic effects. Environmental consequences of this policy were negative. Further continuation of this technical policy is unacceptable.

Creation of new technical means in industries that extract and consume mineral raw materials should be aimed primarily at ensuring environmental safety in the coming years. This means a serious turn in research and development work towards environmental issues. Technical updates are continuously carried out in all mining processes. As technology develops, qualitative change in the human environment gradually occurs.

Examples given allow us to approach the next, most important stage in the development of environmental solutions—assessing state of the environment. The essence of this assessment is to characterize the environmental, social, and economic impacts.

Mineral raw materials have been the main economic resource of Kazakhstan for many decades. Geologists know well that their work in searching for minerals would be completely meaningless and useless if continuation of this work were not in the hands of miners and enrichers (Enviromental 2016).

2 Revealing of Environmental Problems in Territory of Mining and Metallurgical Complexes (MMC) of Kazakhstan

The current state of the mining and metallurgical complex of the Republic of Kazakhstan is characterized by a decrease in production efficiency due to deterioration of mining and geological conditions of developed deposits and need to ensure environmental safety during subsoil development. Mining ecology is addressing this problem.

In accordance with the Development Strategy of Kazakhstan until 2030, main goal of state policy in the field of environmental protection was task of stabilizing level of emissions, discharges, and accumulation of waste, and then reducing them with increasing level of production.

Over many years, mining, and metallurgical complex (MMC) of the Republic of Kazakhstan has accumulated large volumes of overburden rocks, tailings, and slag. Millions of tons of harmful substances are released into atmosphere and hundreds of millions of cubic meters of polluted wastewater are discharged into water basins. All this leads to serious economic, social, and environmental problems.

Kazakhstan, which occupies vast territories of 2725 thousand square kilometers in the center of Eurasia, is characterized by a very vulnerable natural environment. It is represented by many types of landscapes around world—from dry subtropics and sultry deserts to high-mountain tundras and glaciers. There are unique water landscapes, such as the Caspian, Aral, Balkhash, Zaysan, Alakol, Tengiz.

The wealth of mineral resources led to rapid development of mining industry. Old cities were actively expanded and new ones were built. As a result, there was intense pollution of air, water and soil, degradation of flora and fauna, depletion of natural resources, and profound disruption of ecosystems. Negative environmental changes have led to increase in morbidity and mortality in population. Life expectancy in certain regions of the republic is 15–20 years lower than in developed countries.

Subsoil has been and will remain in the XXI century the main source of national wealth for most states, and for our Republic it is also potential basis for becoming one of the 50 most competitive countries in the world. Therefore, one of the main tasks of mining and metallurgical industry of Kazakhstan is to develop effective measures to prevent environmental pollution (Code 2017; Environmental Code 2007).

Let us conduct brief analysis of negative impact of mining and metallurgical complexes on the main elements of environment: atmosphere, land, and water resources.

The main waste-generating industries in Kazakhstan are: coal industry (56–57% of industrial waste, including coal mining, enrichment and agglomeration); ferrous metallurgy (11–12%, including mining, enrichment and processing of iron ore); non-ferrous metallurgy (15–16%, mining of uranium and thorium ores—about 5%); extraction of other minerals—9–10%).

The total area of land occupied by waste exceeds 200 km^2. Every year, amount of industrial waste increases by approximately 1.5 billion tons, and at the same time, the level of use of solid waste is currently low (Kuldeev et al., 2021).

The total amount of harmful substances (according to the State Statistics Committee of the Republic of Kazakhstan) emitted into the atmosphere by stationary sources of pollution was 3097,4 thousand tons in 2021, in 2022—2368,7 thousand tons in 2023—2327,6 thousand tons. Air pollution rises to normal in 15 cities. In Leninogorsk, Ust-Kamenogorsk, Aktyubinsk and Almaty, the maximum permissible level of pollution was exceeded by more than 2,5 times. Increased levels of pollution are observed in Zyryanovsk, Aktau, Temirtau, Taraz, Shymkent, and Petropavlovsk.

The largest gross emissions of pollutants (pollutants) from stationary sources are observed in the Karaganda region (30% or more of emissions in the republic). The main pollutants are dust, sulfur and nitrogen dioxides, hydrocarbons.

The most polluted city is Leninogorsk, where the pollution index increased 1,1 times compared to 2006. The content of formaldehyde increased by 1,5, phenol by 1,61 and nitrogen dioxide by 1.1 times. Industrial emissions are carried out by the zinc plant of OJSC Kazzinc.

The largest air pollutants include thermal power plants, whose emissions amounted to about a million tons. In the Pavlodar-Ekibastuz region, energy enterprises emitted 648,3 thousand tons into the atmosphere. At the same time, amount of ash and slag waste increases, leading to increase territory of ash dumps.

Emissions from non-ferrous and ferrous metallurgy enterprises, such as PA Balkhashmys, JSC Zhezkazgantsvetmet, Ispat-Karmet, determine pollution in the central region of Kazakhstan. In 2012, missions amounted to 834,4 thousand tons.

In the East Kazakhstan region, non-ferrous metallurgy enterprises released 243,9 tons of lead and 447 tons of zinc oxide into the atmosphere in 2020.

Currently, the most difficult environmental situation has developed in regions where large mineral deposits are being developed. The main volumes of harmful substances polluting the atmosphere were formed in the territories of Pavlodar (762,6 thousand tons), Karaganda (664,4 thousand tons), Zhezkazgan (481,5 thousand tons).

The environmental situation in the region of the Ekibastuz Fuel and Energy Complex (EFEC) remains very unfavorable. About 10 thousand hectares have also been disturbed by coal mines, rock dumps and energy facilities, including 8 thousand hectares by mining (Information Bulletin 2022).

Environmental problems with waste are especially acute in the Karaganda region, where several hundred enterprises from many industries are concentrated, which not only produce various products, but also intensively pollute the environment. In the region, atmospheric emissions from stationary sources account for 1/5 of all emissions in Kazakhstan. The most powerful polluters remain Ispat-Karmet OJSC and fuel and energy enterprises, producing 75% of emissions from regional enterprises.

The development of deposits causes disturbance and environment pollution. During extraction of mineral resources, environmentally hazardous zones of influence of waste heaps, dumps, quarries, and boreholes are formed; huge territories of land are alienated, areas significantly exceeding the area of mining workings are withdrawn from circulation.

Territory covered by mining waste in the Karaganda region is more than 60 thousand hectares. In the Karaganda coal basin, over an area of 600 km^2, there are over 80 developing coal deposits. Vast territories are occupied by rock dumps (there are about 40 of them), each of them reaching 1–1,5 km^2. The soils have been turned into technogenic soils.

Today, the Republic contains more than 20 billion tons of waste from enrichment and metallurgical processing (Ministry of Energy 2022; Environmental problems

2013). A significant part of them is a potential raw material for additional extraction of non-ferrous and precious metals, rare and trace elements, and the annual accumulation of industrial waste is about 1 billion tons. 14 million cubic meters are also generated annually. m of household waste, the bulk of which is disposed of in unorganized landfills.

There is practically no system for managing industrial and consumer waste, including hazardous and radioactive waste in the republic. Irrational use of mineral resources leads to their premature depletion and pollution of the environment.

Currently, there is also rapid depletion of natural resources, creating prospects for energy and raw materials hunger. Special aspect of this process is the tendency for those countries that find themselves in the role of suppliers of fuel and raw materials to world market to lose their most important source of income and, consequently, their means of subsistence.

Implementation of mining activities and related manufacturing and energy production inevitably causes a negative impact on the natural environment in areas of deposit development and use of mineral raw materials. Types of negative impacts characteristic of the mining sector, along with waste generation, are deterioration of air quality, water pollution, noise and vibration, significant energy consumption, impact on natural landscapes and biodiversity. Major hazard to water bodies is posed by areas of tailings ponds and rock dumps, where processes of leaching of toxic substances can take place, their seepage through soil and, ultimately, contamination of groundwater.

A separate problem is the disposal of enrichment tailings, which may contain high concentrations of toxic substances (for example, arsenic, lead, cadmium, chromium, nickel, cyanide). At all stages of life cycle of mining enterprises (geological exploration, deposit operation, etc.) emissions occur into the atmosphere. During movement of rocks, small dust particles (suspended solids) enter the atmospheric air. Sources of exhaust gas emissions are mobile sources (cars, special vehicles, and equipment). Pollutants enter atmospheric air as part of emissions during combustion of fuel in stationary installations, during blasting operations and processing of mineral raw materials. In addition to dust, emissions contain heavy metals, carbon monoxide, sulfur dioxide and nitrogen oxides.

As production volumes increase, quality deteriorates and conditions for extraction of solid minerals become more complex, the intensity of environmental pollution increases, which creates growing problems for natural ecosystems and the population in mining areas.

The environmental situation throughout the world is such that environmental deterioration threatens not only the well-being, but also the very life of the population.

The only reasonable alternative to this growth trend is to understand the concept of "sustainable development", which ensures economic and environmental stability in the long term.

3 Environmental Safety Is the Value of the Country

From the first days of gaining its independence, Kazakhstan has been paying special attention to environmental issues. The closure of the world-famous largest nuclear test site in Semey region in the world on the initiative of the President of the country, the encouragement at the international level of programs to preserve the Aral Sea and protect the Caspian Sea, as well as the practical works of Kazakhstan dedicated to protecting the environment from harmful effects, which are of international importance, are one of evidence of this.

The annual messages of the Head of State are aimed at analyzing achievements, determining the direction of the future, and most importantly, at jointly determining the right path to achieving a common great goal.

On the way to a bright future, Kazakhstan, located in the heart of Eurasia, signed the Rio de Janeiro 92 Declaration, the 25 international conventions in the field of ecology and development and ratified the Kyoto Protocol. And, has taken root as active participant in the sustainable development of global processes in Europe and Asia.

Climate change is one of the most pressing problems facing humanity in the new millennium. Dynamics of environmental deterioration that have arisen in connection with climate change, as well as the measures taken by humanity to combat its negative consequences, pose new challenges for Kazakhstan, requiring fundamental measures of influence. In this regard, the UN Basic Convention was ratified in May 1999. In March 1999, the Kyoto Protocol was signed, and a document on its implementation since May 2009 was adopted. The government of Kazakhstan, compared to the 1992 post-Kyoto period, has decided to reduce greenhouse gas emissions by 2020, to 25% by 2050.

To reduce greenhouse gas emissions into the atmosphere, a "green investment" model has been introduced, which provides for strict targeted financing of the project. Finances coming to the republican budget from greenhouse gas emissions will be spent on updating outdated technologies, as well as on introducing new, energy-saving installations and alternative energy sources.

Kazakhstan's choice of the "Green Ecomony" path using energy-efficient and low-emission technologies will reduce the anthropogenic impact on the environment, and will also help to reach a new qualitative level of social and economic development of the country (Decree 2013; Nurpeisova and Kyrgizbaeva 2013).

To implement projects aimed at reducing greenhouse gas emissions, the International Clean Technology Fund, consisting of representatives of the World Bank and the European Bank for Reconstruction and Development, reviewed an application for financing clean technology projects/programs in Kazakhstan in the amount of 1.1 billion US dollars.

The Basel Convention on the Control of Transboundary Movements of Hazardous Wastes and Their Disposal, ratified by the Republic of Kazakhstan (Basel Convention, 1992; Law of the Republic of Kazakhstan, 2003) obliges to take appropriate measures to ensure minimum of waste generation on the territory of the country, taking into account social, technical and economic aspects, which requires not only solving problem of recycling accumulated mining waste, but also technological modernization of existing production facilities.

There are several reasons for addressing the problem of rational use of secondary mineral resources.

Firstly, transition to environmental regulation based on use of modern technologies. The new system for regulating environmentally hazardous industries changes institutional environment for managing production and consumption waste, covering activities of mining enterprises and management of waste from these activities.

Secondly, reason is related to increase in mining, geological and other problems characteristic of the mining sector of the economy. Deterioration of quality of natural mineral raw materials and conditions for mining, depletion of developed deposits leads to fact that quality of natural mineral resources becomes comparable to quality of accumulated technogenic raw materials. As a result, specific component of waste in extracted raw materials increases and conditions are created for the inevitable transition to new level of standards.

Thirdly, problem of integrated use of mining raw materials is still not being solved. Most ore minerals have a multicomponent composition, while deposits are developed, as a rule, using single raw materials. Subsoil users extract the main minerals, and associated minerals and valuable components go to waste from mining and primary processing. As a result, poorly controlled accumulation of waste continues, the rate of accumulation tends to increase, despite efforts made in recent years to increase the use (utilization) of waste from mining and related processing industries.

According to Deloitte analysts (Development trends, 2019), technological development should be a strategic priority for modern mining companies. Companies are encouraged to clearly articulate their values and develop differentiated business models that deliver those values. One trend is increased attention to corporate social responsibility.

Unlike previous years, focus is shifting to ensuring mutual benefits for companies and local communities. Leading mining companies view environmental spending not as a price to pay for regulatory compliance, but rather as a way to gain long-term competitive advantage.

Accumulated waste is, on the one hand, main pollutant of environment, and on the other hand, it represents valuable products that are potentially suitable for processing and reuse to obtain popular building materials.

In more detail, in accordance with Art. 13 of the Code of the Republic of Kazakhstan "On Subsoil and Subsoil Use" TMF containing useful components are classified as minerals. According to the Tax Code, TMF is also classified as "waste" since a fee is charged for its disposal at landfills. After all, the same object cannot be considered both as waste and as a deposit. Here TMF should be considered as secondary raw materials. And if TMF are classified as secondary raw materials and property rights to them are considered, then it will be possible to process them.

This problem of inconsistency of state strategies, along with problems in legislation, was solved by the President of the Republic of Kazakhstan Kassym-Jomart Tokayev, who united the ministries of ecology, geology and natural resources into one Ministry. The Head of State set the new ministry to increase the investment attractiveness of the waste removal and processing sector.

The need for development (processing) of TMF is dictated by the fact that TMF occupies vast territories, easily blown away material, and are a source of increased environmental risk for MMC regions.

In recent years, the Republic of Kazakhstan has adopted several fundamental documents that have determined the policy and long-term strategy in the field of ecology and waste management. Considering the global crisis, as the main pillar of our program, economic means are provided for use to protect nature, optimized to achieve environmental safety in accordance with environmental standards. This direction is carried out in accordance with concept of the transition of the Republic of Kazakhstan to sustainable development in the period 2007–2024, adopted by the Decree of the President of the Republic of November 14, 2006. Within the framework of this program, a complex environmental issue is addressed—ways to solve the issue of recycling and use of waste.

Difficult situation arose with the issue of cleaning waste pipes by municipal treatment facilities of all regional centers and the cities of Astana and Almaty. In many regions of our country, the water sector is also in a difficult situation. Because many water bodies have lost their natural ability of self-purification and self-healing.

For these purposes, the Government approved the "Green Economy" program for 2010–2014. The program is aimed at developing a green economy, reducing anthropogenic impact on the environment and health, restoring, and preserving the natural ecosystem, and developing a management system for improving environmental quality.

Another issue being addressed within the framework of the "Green Economy" program is ensuring the environmental safety of oil extracted from the Caspian Sea. We all know the relevance of this issue in connection with the latest news in the Gulf of Mexico. Such issues arouse the interest of many countries and create the possibility of jointly solving them. In 2007, the Fundamental Convention for the Protection of the Water Area of the Caspian Sea, concluded between the states of the Caspian coast, came into force. This Convention is the first regional convention adopted to determine the main direction of regulation of anthropogenic impact on the Caspian Sea, restoration of biological and other commercial resources of the Caspian Sea, as well as in order to resolve complex issues together with the states of the Caspian coast. The Caspian Environmental Program of the states of the Caspian coast began its work with the support of international organizations and donor countries. Currently, the program office is in Astana.

It should also be noted that there is a need for international support for the initiatives of the Republic of Kazakhstan. During the Pan-European Conferences held in the Palace of Independence in Astana in 2005, our state proposed consideration of issues of adaptation to climate change and the development of a low-carbon economy. And also, at the 5th Ministerial Conference in Seoul, the issues of "green growth" and "environmental safety of economic development" were topical.

During the 6th conference of ministries in Astana in 2005, such important documents as the Astana Initiative, the Declaration of Ministries, and the regional action plan for 2010–2015 were adopted. Kazakhstan has raised the initiative to create a new environmental declaration "green bridge" between Europe and Asia.

Currently, there is a decrease in the level of environmental violations by natural resource users. This was achieved by strengthening measures taken on an administrative and civil basis by environmental inspectorates.

The Head of State, in his Address to the people of Kazakhstan "Allow us build the future together" dated January 28, 2011, instructed the Government to prepare and adopt program based on new principles to provide residents with high-quality drinking water. In some regions of the republic, they use water that does not meet regulatory standards in quality. The country's water users are increasing every year. The main water users are agriculture (75%) and manufacturing (20%). In addition, it should be noted that more than half of the river waters occur in neighboring countries, and there is also a risk of a problem arising in the future due to climate change and transboundary conditions. Our country is rich in groundwater resources, mineral springs, and hydrothermal energy resources.

Currently, bill is being developed to provide industrial production with circular water installations. This will help rationally use water resources. The Ministry ensured planned reconstruction of national hydrometeorological service, that is, it provided 259 hydrometeorological and environmental weather stations, 291 posts, 185 hydrometeorological control stations with weather forecasting products and monitoring (Committee 2022).

Environmental control is carried out over the quality of atmospheric air, water surface, dry lands and the earth's crust. Radiation monitoring is also carried out. Expeditionary research is being carried out in the Caspian and Aral Seas, Lake Balkhash, the Aktau seaport area, the Nura River, the Shchuchye-Borovoye resort area and transboundary rivers of Kazakhstan.

Fulfilling the instructions of the Head of State and Government will make it possible to achieve environmental safety to ensure a favorable environment to improve the quality of life of Kazakhstanis.

However, despite how contradictory this seems, world practice shows the opposite: many countries with natural resources have not been able to cross poverty line due to improper development of these natural resources. All this confirms that the main factor is people, their will, strength, perseverance, knowledge. This is the "golden key", which makes it possible to open wide doors of independence and prosperity.

4 Waste Management in the Republic of Kazakhstan

4.1 About the National Project "Green Kazakhstan"

By Decree of the President of the Republic of Kazakhstan dated October 7, 2021 No. 670, the National Project "Green Kazakhstan" was approved (National Project, 2021).

Head of State K.K. Tokayev named greening the economy and environmental protection as one of the main directions of the economic course, initiating the development of the national project "Green Kazakhstan" for 2021–2025. The national project "Green Kazakhstan" is the basis of the future of Kazakhstan and its healthy Kazakh nation.

The goal of the National Project "Green Kazakhstan" is to create a favorable living environment for population and improve environmental situation, including: improving quality of atmospheric air, effective management of production and consumption waste, increasing the area of green spaces, instilling respect for nature and wildlife, as well as modernization of the environmental consciousness of the population.

The national project "Green Kazakhstan" is a large-scale effort in which 30 organizations from both the public and quasi-public sectors participate. The developer of the national project is the Ministry of Ecology, Geology and Natural Resources of the Republic of Kazakhstan.

The adopted national project "Green Kazakhstan" is designed to improve environmental situation, and therefore create a favorable living environment for all Kazakhstanis. In this regard, the Republic is implementing measures aimed at improving air quality, which means that a lot of work will be done to modernize the production cycles of industrial enterprises of the state.

Currently, Kazakhstan has begun implementing a unique project to create intelligent Smart Grid networks. This project is unique for Kazakhstan and is the first in the CIS. In the future, the government plans to introduce the Smart City concept in the five largest cities of the country—Smart Astana, Smart Almaty, Smart Karaganda, Smart Ontystuk, Smart Aktobe. The successful implementation of this initiative within the framework of the Digital Kazakhstan program will improve the quality of life of the population and the provision of public services.

4.2 Waste Management Is One of the Key Areas of Development Green Kazakhstan in the Republic

By destroying the environment, modern society is destroying its own future. To ensure environmentally sustainable future, we first need:

- monitoring the state of the natural environment;

- regulation and prevention of industrial emissions;
- development and implementation of waste-free and resource-saving technologies.

According to the "Green Kazakhstan" Project, the key direction of economic development in the republic is to improve management system of technogenic mineral formations (TMF).

In recent decades, in many countries of the world, including Kazakhstan, waste management, which includes measures for the management of TMF, as well as measures to prevent their formation, has become particularly relevant.

Today, many foreign countries have experience in developing effective environmental and economic systems of waste-free and resource-saving technology (Gilyazidinova and Ryzhikh, 2009). For Kazakhstan, this experience is useful in terms of using innovative solutions in the field of processing and disposal of ash and slag waste in domestic practice. Problem of recycling slag in construction remains an urgent task, since almost all research is limited to experimental developments. All this raises an urgent need to conduct targeted comprehensive studies of both the slag itself and materials based on it. Therefore, today waste management has become particularly relevant, as one of the key areas for the development of a "green" economy or "green Kazakhstan", i.e. conservation and effective management of ecosystems.

Outstanding interest to Kazakhstan is the experience of Germany, where the Federal Ministry of the Environment developed Waste Prevention Program in 1972.

The Waste Law in Germany was passed in 1972. The Federal Ministry of the Environment has developed a Waste Prevention Program (German Federal 2013). The Circular Economy Waste Management Act also created the opportunity to introduce a nationwide mandatory "single recycling bin". In Germany, all waste is divided into two parts. Citizens pay for the disposal of waste that cannot be recycled. Recyclable waste is removed by producers at their own expense.

The German dual waste collection system includes sorting and processing of certain types of secondary raw materials at enterprises that produce material. Every manufacturing company is interested in processing. There are large waste recycling complexes in Germany. There are about 20 such complexes in Berlin alone.

In the USA, there are cities, for example, San Francisco, where up to 80% of waste is recycled, but there are also cities where much less recyclable materials are collected (De Silguy Cartin 2011).

In Japan, the area is too small to use landfills. 45% of all waste is recycled, 37% of all waste is incinerated, and only 18% is disposed of at a landfill and subsequently stored. Everything that can be recycled is recycled. The Law on Recycling of Packaging and Cans has been in force in Japan since 1995 (Recycling, 2019) and stipulates that every Japanese resident is required to sort and throw away packaging strictly into containers designated for specific materials. Municipal services are engaged in the resorting and storage of waste, and manufacturers turn it into secondary raw materials. It is mandatory to recycle following types of waste: all types of glass containers, plastic bottles, cans, boxes, paper and plastic packaging. Waste utilization in this country is a well-coordinated process in which three participants successfully cooperate: consumers, entrepreneurs, and government agencies.

As a result, Japan is one of the cleanest countries in the world. Industrial waste is used as construction material. Chubu (Nagoya) and Kansai (Osaka) international airports were built on artificial islands created from crushed industrial waste.

Sweden is an example for waste recycling for many European countries. In Sweden, more than 80% of all types of household waste are recycled. Approximately 18% of the remaining waste is incinerated. And only 2% is taken up by the removal of waste to landfills for disposal.

A system of so-called "deposits" is widely used when the price of a product includes the cost of utilization. Garbage removal, disposal or storage is strictly controlled by the state. Huge fine is imposed for illegal waste removal. All recycling enterprises have sensors installed to monitor the maximum permissible concentration of harmful substances. Information from these sensors is constantly supplied to government environmental authorities.

To solve the garbage problem, Singapore has created an effective waste management system based on recycling. According to the Singapore Environment Agency, in 2020 the city produced 21 thousand tons of garbage every day. Of this, 38% is burned, 2% of non-combustible waste is sent to burial grounds on Semakau Island. The remaining 60% is recycled. There are 4 factories burning waste in the city. The volume of incinerated waste is 90% less than the original volume, which greatly facilitates its disposal.

Semakau Island is artificial, it is completely built from garbage. Its construction began in 1999. The island's area is 350 hectares and continues to grow: garbage will be dumped here until 2035. The "garbage island" itself is planted with trees. In world practice, there are various methods of increasing environmental awareness and environmental literacy of the population, as well as attracting them to participate in environmental activities of country, city, and region.

It is known that environmental protection issues are most successfully resolved with the participation of all interested parties in this process. To do this, first, the principle must be respected, reflecting the right of population to access environmental information and to participate in the process of making environmental decisions.

For example, in EU countries the population plays active role in solving problems with household waste. This is greatly facilitated by the organization of widespread publications and informing the population about the benefits of environmental schemes developed by local municipalities, as well as preliminary public opinion polls regarding their implementation. Countries such as Spain, France, Ireland, Italy, Portugal and the UK have successfully introduced separate collection schemes for centralized compost production.

Thus, foreign experience indicates that improving waste management is possible by using effective incentives for all participants in the waste management sector.

Waste management in Kazakhstan is regulated by the Environmental Code (Ecological Code of the Republic of Kazakhstan 2007).

According to the Statistics Committee of the Republic of Kazakhstan, the total amount of hazardous waste generated in the republic in 2022 amounted to

337,4 million tons, or 12% less than in 2021 and 20% less than in 2020. At the same time, the percentage of waste recycling in 2022 was 33%, in 2021–21%, in 2018–6%). The neutralization of hazardous waste at enterprises across the republic amounted to 0,3% (Table 1). Table 2 provides information on the volume of hazardous waste.

As follows from Table 2, the largest amount of hazardous waste was accumulated in the Kostanay region—8348,5 million tons, which is 85,7% of all solid waste accumulated in the republic during the analyzed period.

Waste is stored in special landfills, storage ponds and tailings dumps, constantly forming growing technogenic, dust-intensive landscapes. For example, the

Table 1 Generation, use and disposal of hazardous waste produced in the Republic of Kazakhstan for 2018–2022 (million tons)

Year	Hazardous waste generation	Use of hazardous waste at enterprises	Disposal of hazardous waste, (%)
2018	303,1	19,5	0,6
2019	420,7	45	0,6
2020	355,9	94,7	0,8
2021	382,2	81,8	0,5
2022	337,4	110,1	0,5

Note: compiled by the authors based on the source (Ministry of Energy of the Republic of Kazakhstan 2022)

Table 2 Existence of hazardous production waste at enterprises of the Republic of Kazakhstan in 2018–2022 (thousand tons)

Regions	2018	2019	2020	2021	2022
Kazakhstan	7,864,376	10,101,416	9,601,606	9,789,307	9,742,973
Akmola	10,394	3829	7539	8601	8583
Aktobe	6029	4435	4899	4649	4512
Almaty	5626	62,367	764	764	1270
Atyrau	5725	104	77	18	21
East Kazakhstan	32,242	15,536	95,032	347,196	353,496
West Kazakhstan	5861	237	248	164	158
Zhambyl	15,755	9409	7568	858	6786
Karaganda	11,661	2677	609,937	609,900	607,461
Kostanay	7,175,181	9,418,800	8,344,938	8,263,283	8,348,517
Kyzylorda	5629	16	17	8	8
Mangistau	10,824	2915	540	304	307
Turkistan	10,662	5022	126	27	32
Pavlodar	505,453	498,214	479,242	500,540	346,779
North Kazakhstan	36,721	48,653	30,427	32,358	33,820
Astana city	26,614	19,172	20,254	20,594	21,231
Almaty city	–	–	–	10	–

Note: compiled by the authors based on the source (Ministry 2022)

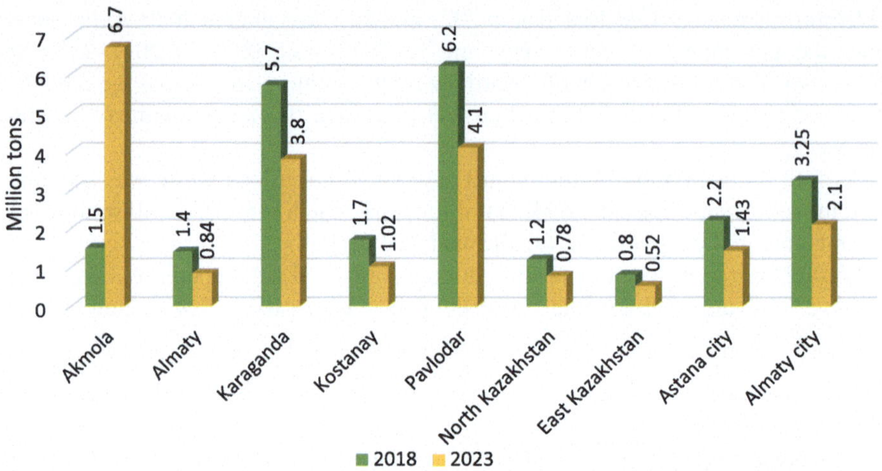

Fig. 3 Generation of ash and slag waste in the regions of the Republic of Kazakhstan

recycling and use of ash and slag waste from power plants in the republic does not exceed 1%, while in Europe this figure averages 60%.

Figure 3 shows volumes of ash and slag waste generation by region of the republic according to the source (Strategic Environmental 2020).

Electric power facilities account for 10% of the annual waste generation. In Kazakhstan, the annual yield of ash and ash and slag mixtures from coal combustion is about 19 million tons; in 2022, more than 300 million tons of waste were accumulated in ash dumps (about 8,5 thousand hectares are used for storage). There is practically no processing of ash and slag waste on an industrial scale. As of 01/01/2022, the share of processed, reused, incinerated ash and slag waste was 0.7%.

It is known that many ash dumps in our country have been in operation for a long time (50–60 years).

In this regard, the resources of almost all ash dumps have been exhausted, so at present it is urgent to expand the capacity of existing ash dumps or build new ash dumps.

In our opinion, utilization problem and processing of ash and slag waste can only be solved through joint efforts of state and enterprises. Technologies for processing ash and slag waste exist. For example, use of ash and slag waste to produce building materials, in laying roads or at other construction sites is economically feasible.

The main reasons for the huge accumulation of waste are insufficiently effective management, lack of economic incentives for development of historical and newly generated waste, and imperfect regulatory framework.

4.3 High Lag for the Green Kazakhstan Project

"Green" does not mean the spring color of awakening nature, "green" is not a reminder of the national currency of the United States, casually called "green" by the people. Here, it seems, everything comes together—it means young. In fact, the "green" economy deserves other definitions, more responsible and balanced in spiritual and moral terms: clean, like a newborn child, wise, like a mature and useful scientist for society, and finally, its main advantage is seen in the fact that it is the most reliable friend and partner of the environment.

Maybe you will say that such miracle in the totality of all above qualities does not exist in nature. Maybe we have really raised the bar for "Green Kazakhstan", but introduced the National Projects of the Republic, such as "Quality and affordable healthcare for every citizen "Healthy Nation", "Quality education", "Educated Nation", "Technological breakthrough through digitalization, science and innovation", "Strong regions are the driver of the country's development", "Sustainable economic growth aimed at improving the well-being of Kazakhstanis", "Green Kazakhstan", "Quality education "Educated nation", and you will understand what these projects represent.

"Green Kazakhstan" basically raises issues of efficient use of natural resources and improving the well-being of citizens through diversifying economy, creating new jobs, including: improving air quality, effective management of production and consumption waste, increasing the area of green spaces, instilling careful attitudes towards nature and wildlife, as well as modernization of ecological consciousness of the population (De Silguy Cartin 2011).

On the territory of the Republic, according to the State Cadastre, about 30 billion tons are stored in dumps, tailings ponds and storage tanks of mining enterprises. Industrial waste, including: 72%—dump rocks of overburden and substandard ores, 20%—dump tailings, 8%—other waste. With annual output of industrial waste of 1 billion tons no more than 100 million tons are usefully used. The rest pollutes the environment, gradually accumulating in it.

According to the bodies of State Control and Supervision of Natural Resources, share of waste used in the republic is 18–20%. For example, in 2017, the percentage of waste utilization was 16%, in 2018–18.98%, and in 2019—20%. However, this figure in the recent past in the industry of the former USSR was 29%. It remains extremely low compared to world practice. In Western Europe (France, Germany, Italy, England) this figure is up to 58%, in North America (USA, Canada)—up to 63%, in Japan—up to 87%, China—up to 37% (Information on the organization, 2022).

Thus, accumulated waste is, on the one hand, the main environmental pollutant, and on the other hand, it represents valuable products that are potentially suitable for processing and reuse to produce commercial products with high added value. The main reason for progressive accumulation of waste in the country is the raw material orientation of our economy.

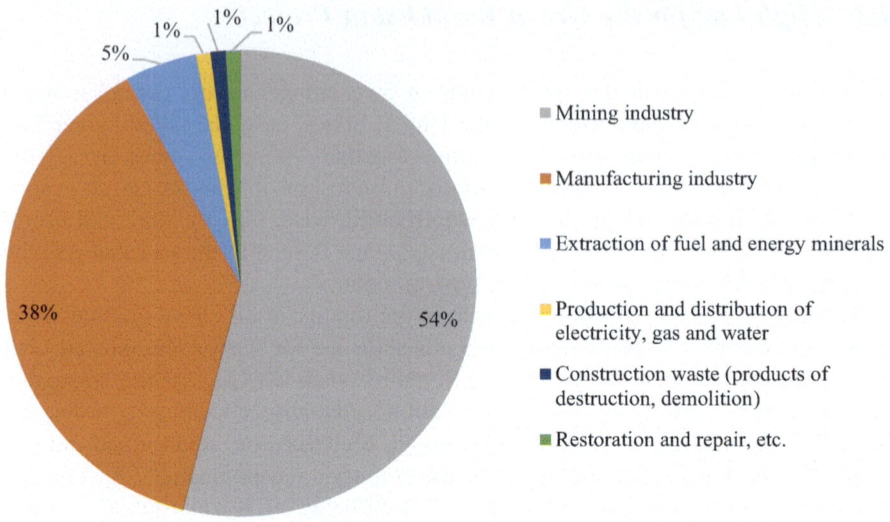

Fig. 4 Main indicators of solid waste accumulated in the Republic by 2023

Currently, more than 22 billion tons of solid waste have been accumulated at 450 landfills in the country (Fig. 4).

Most of the technogenic mineral formations (TMF) are in Karaganda (29,4%), East Kazakhstan (25,7%), Kostanay (17%) and Pavlodar (14,6%) regions. Every year, the republic generates up to 20 million m3 of domestic waste, about one billion tons of industrial waste, including more than 150 million tons of toxic waste (Kuldeev et al., 2025).

Technogenic mineral formations (TMF) are accumulations of mineral substances on the surface of the Earth or in mine workings, formed as a result of their separation from natural massif and storage in the form of waste from mining, processing and metallurgical industries. The constant increase in the volumes of various types of waste generated in mining and processing industries and their storage in storage facilities and experience of using such objects in industry allows us to consider them as sources for obtaining secondary raw materials and building materials.

Purpose of the work is to consider possibility of integrated use of TMF in areas where mining waste is located based on their integrated use as secondary raw materials. Integrated use of raw materials and industrial products progress of metallurgical, mining and construction enterprises is pressing problem not only in Kazakhstan, but also in any economically developed state. As practice has shown, waste from these particular industries is produced in small quantities and poses a serious economic threat. In these conditions, problem of environmentally rational use of industrial waste as secondary raw materials and development of scientific principles in the creation of new technological regulations for production of marketable products from industrial household waste becomes particularly acute. Thus, the introduction of environmentally friendly technology to produce building materials based on waste from mining and metallurgical complex is relevant for Kazakhstan.To develop and implement effective environmental protection measures for waste management, it is necessary to have reliable information about their impact on natural ecosystems: surface, groundwater, air and land disturbance on an industry scale with increasing production volumes. The summarized materials will make it possible to obtain objective information about state of ecosystems in the region under study and to outline priority environmental protection measures, implementation of which will help reduce harmful impact on environment.

It was from this position that the goal was set, idea was justified, and tasks of our research were formulated. KazNITU named after K.I. Satpayev pays increased attention to industrial safety at mines and reducing harmful impact on the environment. Evidence of this is the ongoing research on the project (Project No. BR21882292).

Utilization of technogenic waste from mining and metallurgical industries into construction materials is aimed at solving environmental and social problems in regions with developed mining and metallurgical industries.

Rational organization of waste recycling process in combination with efficient modern equipment makes it possible to obtain products from secondary raw materials at a cost 2–2,5 times lower than for similar products from primary raw materials, with comparable product quality.

5 Conclusions on the First Chapter

In the modern world, the rational use of subsoil resources is one of the principles of sustainable development. The requirement of rationality applies equally to both natural (primary) and technogenic (secondary) mineral resources. Resource potential of waste from mining and related industries accumulated on the territory of the Republic of Kazakhstan is comparable to resource potential of large deposits of solid minerals. At the same time, development and processing of mineral raw materials and growing accumulation of technogenic waste accompanying this activity have a serious negative impact on the ecosystems of the country's mining regions.

Analysis of state statistics shows that for 2020–2023 the average annual growth rate of accumulation of mining waste was about 6%, with average annual growth rate of their formation and use of 7,5 and 9,1%, respectively. According to modern estimates, enterprises of the mining complex of Kazakhstan have accumulated over 50 billion tons of industrial waste and occupy vast territories (more than 200 km^2 of area). Every year the amount of industrial waste increases by approximately 1,5 billion tons, and at the same time, the level of use of man-made mineral formations is currently low.

Group reasons for increase inaccumulation of waste from extraction and processing of minerals are mining-geological, technological, economic, organizational, managerial, and regulatory. The growth of waste processing volumes is hampered by double legal regulation of the use of mining waste, which makes it difficult to actively involve technogenic mineral resources in economic circulation. Measures for economic incentives for processing mining waste are not effective enough; the existing procedure for obtaining use rights to develop technogenic deposits and the current taxation demotivate investors and interested small and medium-sized mining businesses.

"The wealth of the earth's bowels is the well-being of the entire future generation... We are all obliged to live and work as if we do not have such wealth in our hands," said the Head of State, which means preserving existing wealth, efficient use, ensuring environmental safety and living in a bright the future—everything is in our hands.

The integrated use of mineral resources and industrial waste from mining and metallurgical complex is an urgent problem not only in Kazakhstan, but also in countries with developed economies. It is especially important to consider and realize waste as secondary raw materials, that is, production of commercial products from waste.

To summarize, we would like to say that care, caring attitude, respect for the natural, wide land of Kazakhstan is necessary both for ourselves and for sustainable step into a bright future. This is source of wealth, life, basis of the national concept and idea. This means that preserving and nurturing Mother Nature is our national duty.

References

Basel Convention on the Control of Transboundary Movements of Hazardous Wastes and Their Disposal is an international treaty governing the international movement and disposal of hazardous waste. -Basel. 05/05/1992 (In Russian)

Club of Rome (1972). The Limits to growth; a report for the Club of Rome's project on the predicament of mankind. New York, Universe Books

Code on subsoil and subsoil use Republic of Kazakhstan dated 27.12.2017. 125-VI RK (In Russian)

Committee for Environmental Regulation and Control of the Ministry of Ecology, Geology and Natural Resources of the Republic of Kazakhstan, Electronic resource, 02.03.2022. (In Russian)

Deloitte: Development trends - 2019. 10 key factors influencing the future of the mining sector. pp. 29–32. https://www2.deloitte.com/ru/ru/pages/energy-and-resources/articles/tracking-the-trends.html (date of access 15.10.2019)

De Silguy Cartin. History of garbage. Text, 2011. (In Russian)

Decree of the President of the Republic of Kazakhstan "On the Concept for the transition of the Republic of Kazakhstan to a "green economy" dated 30.05.2013, № 577.(In Russian)

Environmental Code of the Republic of Kazakhstan. - Astana, 2007.(№212-Sh).(In Russian)

Environmental and industrial safety of subsoil development (monograph under the general editorship of M.B. Nurpeisova). Almaty: KazNTU, 2016. 430 p. (In Russian)

Environmental problems of mining and metallurgical complexes. Integrated processing of mineral raw materials of Kazakhstan. Almaty: Foliant, 2013, volume 8. 271 p. (In Russian)

Galiev SS (2015) Scientific and innovative potential of the mining and metallurgical complex of Kazakhstan. Materials of international scientific-practical.conf.-Almaty, 87–97. (In Russian)

Gilyazidinova NV, Ryzhikh NI (2009) Resource-saving technologies in construction and operation [Electronic resource]. (In Russian)

German Federal Ministry of the Environment (2013) Waste Prevention Program. (In Russian)

Information environmental bulletin of the Republic of Kazakhstan (2022) Natural resources and ecology of Kazakhstan at the turn of the 21st century. Special issue, Almaty. 13 p. (In Russian)

Information on the organization of waste management in the regions of the Republic of Kazakhstan. Astana, 2022. 154 p. http://ecogosfond.kz/wp-content/uploads/2022/11/Informacionnyj-obzor-po-vedeniju-gosudarstvennogo-kadastra-othodov-za-2022_compressed.pdf

Kuldeev EI, Nurlybayev RY, Orynbekov YS, Bek AA, Ashimova AA (2025) Study of ash and slag waste for their use as secondary resources. International Scientific Conference «Prospects for the application of digital technologies in rational and safe subsoil use», Tashkent: Tashkent State Technical University, pp 287–291

Kuldeev EI, Nurpeisova MB, Kyrgizbaeva GM (2021) Subsoil development and environmental safety. Deutsschland LAP LAMBERT, 234 p. (In Russian)

Law of the Republic of Kazakhstan dated February 10, 2003 №389 "On the accession of the Republic of Kazakhstan to the Basel Convention on the Control of Transboundary Movements of Hazardous Wastes and Their Disposal". (In Russian)

Meadows (1972) Limits to growth. Translation from English. (In Russian)

Melnikov NN (2003) Environmental problems of the 21st century and the development of natural resources. In Subsoil development and environmental problems - a look into the 21st century. M.: IPKON RAS. 26–45 pp. (In Russian)

Ministry of Energy of the Republic of Kazakhstan (2022) Official Internet resource [Electronic resource]. http://energo.gov.kz/index.php?id=2960 (date of access 24.09.2022)

National project "Green Kazakhstan" Resolution of the Government of the Republic of Kazakhstan dated October 12, 2021 № 73. (In Russian)

Nurpeisova MB, Kyrgizbaeva GM (2013) State and prospects of the mineral resource base of Kazakhstan. Materials of the 10th international conference. "Problems of subsoil development in the 21st century through the eyes of young people. M.: IPKON RAS. 116–119 pp. (In Russian)

Project No. BR21882292 "Integrated development of sustainable construction industries: innovative technologies, optimization of production, effective use of resources and creation of technological park" with the support of the Science Committee of the Ministry of Science and Higher Education of the Republic of Kazakhstan. (In Russian)

Recycling of household waste in Japan [Electronic resource]. 2019. http://ecomg.ru/about/article/20 (date of access 26.09.2019). (In Russian)

Strategic environmental assessment of the Concept for the development of the fuel and energy complex of the Republic of Kazakhstan until 2030. EU/UNDP/UNECE Environmental Report. 2020. (In Russian)

Subsoil use 2000–2012: reserves, production, investments. -Almaty, Kazakhstan. AMM 2012. 36–40 p. (In Russian)

Yoda. Trilemma threatening the world, M., MGF, Knowledge, 1999. 186 p. (In Russian)

Prospects of Use of Mining Waste for Production of Construction Materials

Abstract Analysis of trends in the field of formation and use of waste from mining and beneficiation of minerals is carried out. Reasons for the growth of waste accumulation are identified.

Problem of waste involvement in economic circulation is considered, an assessment of the possibility of its solution is given taking into account the contradictions of the mining and environmental legislation of Kazakhstan. Measures are proposed to stimulate the involvement of waste in economic circulation by introducing differentiated rates for the tax on mineral extraction during ore processing, creating a cadastral system and automated databases, and developing a waste management system.

1 Rational Use of Subsoil Resources Is One of Principles of Sustainable Development

Topic of rational use of subsoil resources cannot be called new. Issues of rational subsoil use, one way or another affecting mining and enrichment waste, have been studied by more than one generation of Russian (and previously Soviet) scientists. The peculiarity of the author's approach to the issue is that monograph considers gas and mineral resources in the aspect of their dual nature: as secondary mineral resources and as objects of negative impact in conditions of greening industrial production.

Academician K.I. Satpayev paid special attention to the issue of using rich subsoil resources in his time as a mining engineer-geologist. He was one of the first in the USSR to draw attention to the importance and necessity of rational use of mineral resources. Back in 1962, he noted that the rational use of subsoil resources is the most pressing task in the non-ferrous metallurgy of the republic and country (Satpayev 1962).

In 2024, the entire scientific community and the entire country celebrate the 125th anniversary of the birth of the great son of the Kazakh land, an outstanding

M. Nurpeissova et al., *Development of Technology for Processing of MMC Waste*, SpringerBriefs in Earth Sciences, https://doi.org/10.1007/978-3-032-12976-5_2

scientist of the twentieth century, academician of the USSR Academy of Sciences Kanysh Imantaevich Satpayev.

Kanysh Imantaevich Satpayev is the first president of the Academy of Sciences of the Kazakh SSR, the founder of many areas of science, the creator of the scientific school of metallogeny, the discoverer of the most important mineral deposits of Kazakhstan, a generally recognized organizer of world-class production and science. Academician K.I. Satpayev had phenomenal abilities to influence others, look into the future and was a truly cosmic personality.

Therefore, it is no coincidence that on his birthday, April 12, 1961, in this native steppe, where he spent his young creative years, the first person on the planet, cosmonaut Yu.A. Gagarin, flew into space. This epoch-making event confirmed the superiority of the education system, science, technology and the triumph of the victory of the human mind.

It is very symbolic that April 12, the birthday of the Great Scientist of our country, has been designated as Science Workers' Day by Decree of the President of the Republic of Kazakhstan since 1998.

K.I. Satpayev was and remains the father of Kazakhstani science. Currently, the Code of the Republic of Kazakhstan "On Subsoil and Processing of Mineral Raw Materials" considers the rational use of subsoil in an inextricable connection. The key requirement is to ensure the most complete extraction from the subsoil of reserves of the main and co-occurring minerals and associated components. The principle of rational use and protection of subsoil in the interests of present and future generations is one of the main most important principles of sustainable development.

Rational is the use of subsoil, which involves, along with the complete and comprehensive extraction of minerals contained in the subsoil, "the most efficient use of other subsoil resources using the best modern technologies for development or processing of their subsequent operation, ensuring the minimization of harmful effects on other components natural environment.

Developed countries of the world have long ago come to understand the need to concentrate responsibility for the management of production and consumption waste at the national level. Our country still does not have a unified subsoil waste management system.

Kazakhstan's subsoil legislation prescribes that use of mining waste is considered part of the mineral extraction process, which equates mining and processing waste to natural subsoil resources. This aspect does not consider features of recycled mineral raw materials as a potential resource and is a demotivating factor in relation to the processing of stale waste.

As a result of the administrative equalization of procedure for obtaining subsoil use rights in relation to primary and secondary mineral resources, the latter become and will remain investment unattractive for subsoil users and investors, and for small mining businesses interested in processing technogenic raw materials, the barriers to entry into the industry turn out to be practically insurmountable.

Not the least role in creating barriers to the rational use of accumulated and continuing to accumulate waste is played by the lack of normatively established concepts of "secondary mineral resources", "technogenic deposit" and others.

Modern mining production, as before, needs an early transition to innovative, high-tech development. The impetus for such development can be given by the introduction of the concept of modern technologies, which combines the technological, environmental, and economic aspects of the methods and equipment used into practically implementable complexes with known efficiency. The same effect can be obtained in relation to increasing the complexity of the use of natural or technogenic mineral raw materials.

Traditionally, the use of mining waste is considered as a factor that ensures the expansion of the mineral resource base, overcoming shortage of natural mineral raw materials, saving money on waste disposal, reducing the need for investment for the development of new deposits, preventing and reducing environmental damage, and obtaining additional social effects. However, not all waste from mining and processing of minerals can be considered as mineral resources suitable for industrial development.

2 The Role of Environmental and Industrial Safety During Subsoil Development

Based on importance of mineral resource complex for economy of the Republic of Kazakhstan, strategy for its development should be implemented throughout country and, therefore, be state strategy. Goal of this state strategy is to select optimal solutions, implementation of which will satisfy Kazakhstan's needs for minerals and ensure its stable and safe position in the global mineral market.

To achieve this goal within the framework of such a strategy, it is necessary to obtain answers to the following questions (Code 2017):

1. What types of mineral resources are it advisable to mine directly on the territory of the Republic of Kazakhstan?
2. What volumes of mineral extraction, and in what time frame, must be ensured?
3. What mineral deposits, including man-made ones, should be exploited?
4. What technological solutions should be used in the extraction and processing of minerals and mineral raw materials?
5. What technology should be used in the extraction and enrichment of minerals and raw materials?
6. What human resources must be in demand, both in quantity and quality?
7. To what extent should financial resources be mobilized (public and private investment) to implement this strategy.

It is extremely difficult to get answers to these questions, since the economy of Kazakhstan uses more than 100 types of minerals, which can be of different quality, obtained from different sources, extracted and processed using different equipment and technology, and used in different ways in different sectors of the national economy. In general, we are faced with many options for the development of the system of providing the country with mineral raw materials and energy resources. Consequently, a methodology must be developed and applied to minimize these options to a level that allows the necessary feasibility study to be carried out.

Currently, two-stage model has actually emerged in the economy of our country: the public sector and the private sector. There is real uncertainty in the future development of these sectors, which makes it difficult to draw up scientifically based strategic forecasts.

In Kazakhstan, it is necessary to implement such a principle of rational subsoil use as an organic combination of market mechanisms for regulating and supporting efficient resource consumption. The system of state regulation and support for rational and safe subsoil use should include, at a minimum, the following subsystems:

- legal support;
- financial and credit support;
- infrastructure and information support;
- insurance against emergencies and natural disasters.

Combination of strategic and operational current measures can lead to a fundamentally new quality of solving the problem of rational and safe subsoil use, in particular, increasing the economic efficiency of using the state subsoil fund.

Strategy for the development of the country's mineral resource base and mining industry depends on the state of the economy of modern Kazakhstan, what model of its development and what actions of the state should correspond to this.

There are three known models for constructing modern market systems (Nakipelov 1999):

- neoliberal model;
- model of neosteitism;
- model of neo-corporatism.

It is quite obvious that the modern Republic of Kazakhstan, given its economy, cannot immediately move to the "neoliberal" model, which gives the greatest increase in GDP. Consequently, at this transitional stage it is necessary to implement the strategy of a "neostativist" model, which, given the significant role of the mineral resource complex (MSC) in the economy of Kazakhstan, should ensure the primary role of the state in choosing the priorities for the development of the MRC.

Modernization and development of MSK is the most realistic and shortest way to revive the economy and its transition in the future to a "neoliberal" model.

> At the same time, it must be taken into account that the scale and depth of the negative impact of mining on the environment causes great concern to society. Even poets write with excitement:"There is less and less surrounding nature,
> More and more environment."

It has been established that the impact of mining on the environment must be identified a priori in order to have a reserve of time to develop the most technologically and economically effective methods for eliminating and minimizing this impact.

When developing technological processes for mining production, priority should be given to those solutions that ensure the absence or minimum generation of production waste (Fig. 1).

The idea of creating waste-free (ecologized) subsoil use involves the development and implementation of methods and means of their organizational and economic support, allowing modern mining production to fit into the natural geochemical cycle, thereby turning it into a geochemically closed system.

Closed green mining production is based on the following principles:

- minimum losses of matter and energy at the stages of their removal from the natural system and subsequent use in mining;
- maximum use of mining waste in other economic systems and to restore the disturbed ecological balance of the natural system.

These principles should be taken into account when developing the scientific basis for an environmental strategy for the development of the mineral resource base and mining industry in the Republic of Kazakhstan.

All methods of mining are characterized by an impact on the biosphere, polluting almost all its elements, including the subsoil.

Subsoil is the object and operational basis of mining production and is subject to the greatest impact. Since subsoil belongs to elements of the biosphere that do not have the ability to naturally renew in the foreseeable future, their protection should include ensuring scientifically sound and economically justified completeness and complexity of use.

The impact of mining on the biosphere is manifested in various sectors of the national economy and is of great social and economic importance.

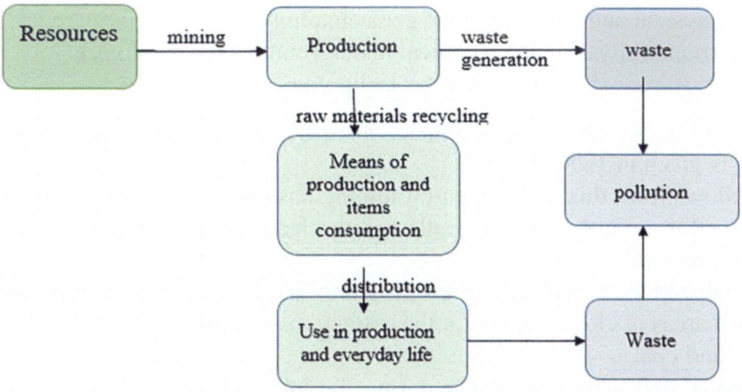

Fig. 1 Conversion of subsoil resources into environmental pollution

Table 1 Comparative assessment of the impact of various types of industrial production on the environment

	Impact of industries on elements of the biosphere						
		Water pool		Earthly surface			
Industry	Air pool	Superficial water	Underground water	Soil cover	Landscape	Flora, fauna	Subsoil
Chemical and * petrochemical	S_t	S_t	M_e	M_e	Mi	M_e	Mi
Metallurgical	S_t	S_t	Mi	M_e	Mi	M_e	O
Fuel and energy	S_t	S_t	Mi	Mi	Mi	Mi	O
Construction	Mi	Mi	Mi	M_e	M	Mi	Mi
Transport	M_e	M_e	Mi	Mi	Mi	Mi	O
Mining	M_e	S_t	S_t	S_t	S_t	M_e	S_t

Note. O - no impact, Mi - minor impact, M_e - medium impact, S_t - strong impact

At present, it is not possible to give a comparative quantitative assessment of the impact of subsoil crawling and other types of human activities on the environment, since there is no scientific and methodological basis for such a comparison. The various private criteria used do not allow us to obtain an unambiguous answer to this question.

In Table 1, A comparative qualitative assessment of the environmental impact of certain types of industrial production is given (Pevzner 2008).

As can be seen from this table, mining has the widest impact on the biosphere, affecting almost all its elements.

There are a number of examples to classify the impact of mining on the environment. These classifications, along with their advantages, have significant disadvantages and do not reflect all the features of the impact of subsoil use objects on the environment. It is considered appropriate to classify the impact of subsoil use objects on the environment according to individual elements of the biosphere.

At the present stage of development of science and technology, solid mineral deposits are developed mainly in three ways:

- open (physical and technical open geotechnology);
- underground (physical and technical underground geotechnology);
- borehole (physical and technical geotechnology).

A comparative assessment of the impact of various methods of mining on the environment is given in Table 2.

As follows from this table, open-pit mining has the greatest impact on the environment, while mining through wells has the least impact, i.e. geotechnological (borehole) method.

We analyzed the results of various programs and projects for environmental restoration in areas of closed mines for the extraction of uranium, ore and non-metallic minerals and coal.

Activities that were carried out during the implementation of these programs included: inventory of areas of underground and open-air mining complexes,

Table 2 Assessing the environmental impact of mining methods (Pevzner 2008)

Mining method	level of impact of mining on elements of the biosphere				
	Water pool	Air pool	Flora, fauna	Earth, soil	Subsoil
Open	S_t	S_t	S_t	S_t	S_t
Underground	M_e	O	M_i	M_e	S_t
Borehole	M_e	M_i	O	M_i	S_t

Note. O - no impact, M_i - minor impact, M_e - medium impact, S_t - strong impact

creation, management and visualization of a database, assessment of environmental impact, forecasts of future changes in this impact, classification and prioritization of individual and group underground and above-ground objects complexes, development of technical solutions for the liquidation or conservation of objects with subsequent technical and biological reclamation of disturbed areas, management and purification of mine and drainage waters and simultaneous environmental monitoring.

For example, mining enterprises annually extract over 200 million tons of minerals and host rocks from the depths of the Republic of Kazakhstan. The extraction and processing of such large volumes of mineral resources have a negative impact on the environment due to the generation of mining waste. Subsoil use waste is stored off-balance ore, overburden rock, intermediate waste from the process of ore preparation and hydrometallurgy, waste tailings from the flotation and sorption process, slag from the pyrometallurgy process, liquid cyanide containing waste, which is placed on internal and external dumps, tailings and ash and slag storage facilities and other storage facilities.

Mining in Kazakhstan continues to have a negative impact on environmental components.

During the extraction of mineral deposits, huge amounts of hazardous waste are produced, which detonate in the dumps of mines and mines, as well as in the tailings of processing plants and hydrometallurgical plants.

Solution to environmental problems in the areas of mines and processing plants cannot be delayed. Intervention at the highest levels of government is urgently needed. Modern society is seriously concerned about solving problems of mining ecology and industrial safety, on which the well-being of the current and future generations involved in subsoil use depends (Begentayev 2024).

Thus, subsoil use objects are the main cause of negative impacts on the environment and mining safety. Neutralization of such impacts during the modernization of the mining industry will play a positive role in the formation of a strategy for the development of the mineral resource base; it requires, on the one hand, a certain organization of production and, on the other hand, the use of a method of rational development of subsoil resources, taking into account the problems of mining ecology and industrial safety, which is positive will influence the choice of priorities for the development of the mining industry.

3 Main Sources of Waste Generation

Negative impact of industry is expressed in the impact on specific parts of nature and on the biosphere in whole from processes of extraction and processing of natural resources. Production and consumption waste are sources of anthropogenic environmental pollution on global scale and arise as inevitable result of consumer attitudes and unacceptably low resource utilization rates.

For example, in the USSR, non-ferrous metallurgy consumed about 2 billion tons of rocks per year, and commercial output accounted for 1%. In the Russian Federation, one way or another, 90–95%, or from 80 billion to 120 billion tons, end up as waste. More than a billion of them are toxic. Every year, the area occupied by waste increases by 250 thousand hectares. The main suppliers of waste are the mining, chemical, metallurgical, fuel and energy industries.

Production waste is the remains of materials, raw materials, semi-finished products formed during the manufacturing process and which have lost, in whole or in part, their useful physical properties.

Industrial waste can be considered products formed as a result of physical and chemical processing of raw materials, mining and enrichment of minerals, the production of which is not the purpose of this production. Consumer waste—machines, tools, household products unsuitable for further use for their intended purpose and written off in the prescribed manner.

Depending on the possibility of use, distinction is made between recyclable and non-recyclable waste. For the former, there is technology for processing and inclusion in economic circulation, for the latter it is currently not available. Classification of industrial waste (Esmesov and Bekbasarov 2009). Industrial wastes are often chemically heterogeneous, complex polycomponent mixtures of substances with various chemical and physical properties, presenting toxic, chemical, biological, corrosive, fire and explosion hazards.

There is a classification of waste according to its chemical nature, technological characteristics of formation, and possibility of further processing and use. In our country, harmful substances are characterized by four hazard classes, which determine costs of processing and disposal:

1. Extremely dangerous. Waste containing mercury and its compounds, including sublimate (HgCl2), potassium chromate and cyanide, antimony compounds, including SbCl3—antimony trichloride, benzo-a-pyrene, etc.
2. Highly dangerous. Waste containing copper chloride, containing copper sulfate, copper oxalate, antimony trioxide, lead compounds.
3. Moderately dangerous. Waste, lead oxides (PbO, PbO2, Pb3O4), nickel chloride, carbon tetrachloride.
4. Low risk. Waste containing magnesium sulfate, phosphates, zinc compounds, waste from mineral processing by flotation using amines.

Belonging to groups is determined by the industrial waste classifier, by calculation if hygienic parameters of substance are known (for example, maximum permissible concentrations) and experimentally. Waste of all classes is divided into solid, liquid, dust or gaseous.

Number of waste classifications are known in modern scientific and technical literature N.F. Reimers (1994) divides waste into industrial, agricultural, construction and municipal. M.E. Pevzner (2008) considers only MMC waste and divides it into solid, liquid and gaseous. Ch. Mantell (Solid waste 1979) identifies 6 classes of waste: agricultural, livestock, metallurgical, construction, chemical and municipal. The works of Z.A. Estemesov and T.K. Sultanbekov (2001) proposed classification of urban construction waste into 13 groups.

The waste classifier of the Republic of Kazakhstan (Estemesov and Bekbasarov 2004) consists of 8 blocks of multi-digit codes and 20 types of different waste.

Code groups (blocks) of the Classifier:

- name (N);
- reasons for transferring the material to waste (Q);
- aggregation state of waste (W);
- hazard identifier of waste components (C);
- properties that determine hazard of waste (H);
- implemented waste management method (D, R);
- the main type of activity as a result of which waste was generated (A);
- industrial waste hazard level (G, A, R).

This Waste Classifier is applied document and is not normative legal act (NLA). At the same time, the Ministry uses it as a legal document to determine level of hazard and coding of waste.

In this work, to understand mechanism of occurrence, control, and management, as well as to eliminate shortcomings noted above, classification of production and consumption waste was made according to physical condition, by sources, chemical and morphological composition, as well as by industries and regions (Fig. 2).

This classification includes the following types of waste:

Fig. 2 Classification of waste by characteristics physical condition and education

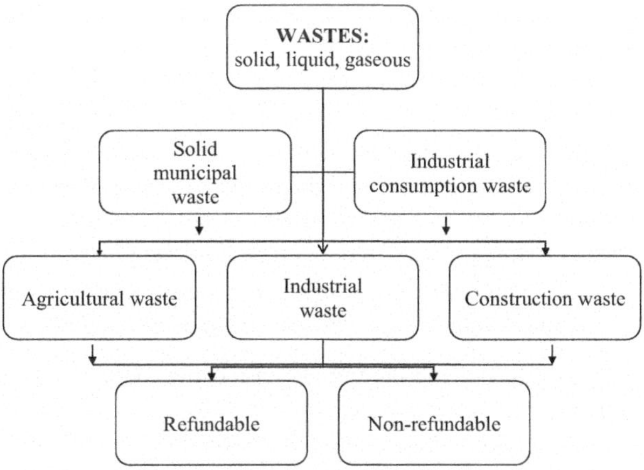

1. mining industry;
2. agriculture, gardening, hunting;
3. forestry, pulp and paper and woodworking industries;
4. leather and textile industries;
5. oil and gas processing;
6. basic chemistry;
7. chemistry of organic synthesis;
8. production, processing, supply and use of paints, varnishes, sealing materials, adhesives, printing inks and enamels;
9. photographic industry;
10. inorganic waste from thermal processes;
11. inorganic waste with metals from metal processing;
12. machine and mechanical processes;
13. mineral and synthetic oil waste;
14. from using substances as solvents;
15. contaminated packaging, contaminated adsorbents and filters;
16. industrial waste not otherwise defined in the classifier;
17. commercial construction and demolition of buildings;
18. medical and veterinary services and research;
19. processing and water production;
20. household waste and similar trade waste

4 Main Reasons for Increase in Accumulations of Mining Waste and Primary Processing of Minerals

Waste is the remains of raw materials, materials, nonstandard and byproducts, used finished products that have been used and have lost their original consumer qualities, placed in certain places with subsequent mandatory use, processing or disposal. They are in solid aggregate state (some gaseous and liquid wastes can pass into the solid phase, for example in filters or settling tanks) (Bykhovsky and Sporykhina 2011).

Consumer waste is products and materials that have lost their consumer properties as a result of physical (material) or moral wear and tear.

Industrial (production) waste is the remains of raw materials, materials, semi-finished products generated during production of products or work and which have lost completely or partially their original consumer properties. They can be solid (waste metals, plastics, wood, etc.), liquid (industrial wastewater, waste organic solvents, etc.) and gaseous (emissions from industrial furnaces, vehicles, etc.).

The main causes of waste are (Bekbasarov 2008):

firstly, irrational management, which has become the norm for many enterprises using old technologies;

secondly, outdated regulatory framework.

thirdly, ineffective control of central and local environmental and health authorities and other sectoral government bodies;

fourthly, absence of special law regulating relations in the field of waste management. The ministry's attempt to solve waste management problems by introducing amendments and additions to the Law on Environmental Protection (2007) is unrealistic. It is impossible to solve such a huge problem by amending several articles of the law.

It is known that in the environmental aspect our republic is one of the most polluted countries in the world. One of the main sources of pollution is municipal solid waste (MSW).

Most cities with a population of over 100 thousand inhabitants, especially megacities like Almaty, are experiencing an environmental crisis due to production and consumption waste.

The high concentration of population in megacities leads to increase in anthropogenic load on environment, which consists in increasing the production and consumption of various energy values. This leads to environmental pollution and disruption of ecology of the urban landscape. Disposal of waste generated by the city to landfills leads to pollution and irrational use of land, the atmosphere, surface and ground waters, increased transportation costs and irretrievable loss of valuable materials and substances.

The solution to this problem is currently urgent due to the manifold increase in the volume of solid waste. In this regard, the end of the twentieth century was marked by the holding of several International Environmental Congresses in the following countries: Brazil, Canada, Japan, Portugal, Sweden, and Russia. Several laws and codes have been adopted in Kazakhstan, which confirms the global nature of environmental problems associated with the lithosphere, atmosphere, hydrosphere and biosphere. The scientific foundations and practical recommendations for environmentally sound management of industrial waste are formulated in the works of scientists from Kazakhstan, the CIS countries and abroad: A.K. Adryshev, O.A. Baikonurov, M.Zh. Z.A. Estemesov, B.N. Laskorin, M.E. Pevzner, K. Seitzhanov, K.N. Trubetskoy, N.N. Chaplygin and many others.

However, despite certain achievements in this area, the problems of organizing burial, storage, neutralization and disposal of waste occupy a leading place among environmental problems. Insufficient attention is paid to the processing and disposal of mining waste, construction and municipal waste.Over many years, the mining and metallurgical complex of the Republic of Kazakhstan has accumulated large volumes of overburden waste, enrichment tailings, and slag. Millions of tons of harmful substances are released into the atmosphere and hundreds of millions of cubic meters of polluted wastewater are discharged into water basins. All this leads to serious economic, social and environmental problems (Fig. 3).

Scale of the impact of industrial waste on environment is comparable to geological processes. Large-tonnage industrial wastes are accumulations of waste from the mining, metallurgical, energy and chemical industries containing useful components or minerals. The bulk of mining waste in Kazakhstan is generated in mining

Fig. 3 General view of industrial waste

enterprises (73%), processing plants (25%) and metallurgical plants (2%). According to modern estimates, the enterprises of the mining complex of Kazakhstan have accumulated over 50 billion tons of industrial waste and occupy vast territories (more than 200 km2 of area). Every year the amount of industrial waste increases by approximately 1,5 billion tons, and at the same time, the level of use of solid waste is currently low. The total reserves and areas occupied by tailings dumps within the regions are given in Table 3.

Moreover, the largest reserves are concentrated in tailings dumps. The need to involve enrichment tailings in production is dictated by the following circumstances:

- service life of tailings dumps is limited; the filling of many has already been completed or will end in the coming years;
- tailings occupy vast territories and, because they are finely dispersed and easily blown away material, are source of increased environmental risk for regions where mining and processing complexes operate.

Since enrichment waste is finely ground product that does not require additional grinding before use, this reduces economic costs. In addition, the process of ore enrichment ensures the homogeneity of the material both in chemical and mineralogical composition. The total reserves of mining and processing waste of some large mining enterprises in Kazakhstan are given in Table 4.

According to the bodies of State Control and Supervision of Natural Resources, the share of waste used in the republic is currently 18–20%. It remains extremely low compared to world practice. In Western Europe (France, Germany, Italy, England) this figure is up to 58%, in North America (USA, Canada)—up to 63%, in Japan—up to 87%, China—up to 37% (Saparova et al. 2016).

Accumulated waste is, on the one hand, the main pollutant of environment, and on the other hand, it represents valuable products that are potentially suitable for processing and reuse to obtain popular building materials.

The need for development (processing) of technogenic mineral formations is dictated by the fact that they occupy vast territories, easily blown away material, and are a source of increased environmental risk for the mining and metals industry regions. The constant increase in the volumes of various types of waste generated in

Table 3 Processing waste by regions of Kazakhstan

Regions	Quantity	Reserves, thousand tons	Area km^2
Akmola	11	76,834,50	12,30
Aktobe	8	30,675,30	6,30
Almaty	5	47,914,90	2,99
East Kazakhstan	39	887,914,57	19,57
Zhambulskaya	6	44,188,93	1,58
Karaganda	37	2,809,342,13	89,20
Kostanayskaya	4	611,101,70	27,45
Pavlodar	2	8770,86	1,23
South Kazakhstan	5	142,355,30	3,52

Table 4 Reserves of technogenic waste at large mining enterprises

	Reserves, thousand tons	
Name of the enterprise	Technogenic dumps	CP dumps
JSC “Achpolitmetal”		142,570,1
Belogorsky GOK	24,406,0	10,067,8
Donskoy GOK	81,447,7	38,280,4
Kazakhmys LLP	973,114,7	1,674,691,5
Zhairemsky GOK	6354,8	3188,8
Tekeli mining and processing plant	15,723,9	40,360,5
JSC “Kazzinc”		373,147,1
Zhezdinsky GOK	89,7	3173,2
JSC “Kostanay minerals”		2038,3

the mining and processing industries and their storage in storage facilities and the experience of using such objects in industry allows us to consider them as sources for obtaining secondary raw materials and building materials.

Continued trend towards increasing accumulation of mining and processing waste is a consequence of an increase in their formation and insufficiently intensive processing (utilization), which is due to several interrelated factors: mining-geological, mining-technical, economic, organizational, managerial, legal and institutional.

The increase in waste generation during mining is associated with an increase in production volumes, deterioration of mining and geological conditions for mining operations, low efficiency of the technologies used, and the existing practice of providing deposits for development.

Deterioration of mining and geological conditions for mining operations, particularly decrease in the depth of development of deposits using open-pit mining, leads to an increase in the volumes of extracted waste rocks and low-grade mineral resources that are sent to dumps. It is no coincidence that, for example, the greatest increase in waste is observed in the extraction of fuel and energy minerals, primarily coal, the share of open-pit mining of which exceeds 75%.

In addition, domestic mining production is characterized by high level of losses at the stages of production and processing. For example, when implementing all new potash projects, it is planned to use only the traditional (mine) mining method, in which on average up to 60% of the ore is lost in the ground.

The increase in the volume of accumulated waste is facilitated by lack of a waste management subsystem in state management system of subsoil use, as well as the lack of legal and technical regulations for the formation of technogenic deposits, the establishment of property rights, and a unified accounting system. Mechanism for limiting the extraction of mineral resources does not work if there are sources of waste of this type in the region.

The other side of the problem of mining waste accumulation is the possibility of their further processing. It should be considered that mining production is a combination of complex, technologically and organizationally related processes. Commonly processing of stored waste is impossible without stopping the associated production and disrupting the technological cycle, therefore, the development of such waste can only be carried out when their dumps have already been formed (closed).

One of the reasons holding back the processing of previously formed stale waste, is deterioration over time of their qualitative and quantitative composition. During long-term storage, physical and mechanical properties of rocks also change. Thus, studies of overburden rocks from the dumps of the Zhezkazgan copper ore deposit, used in the Ulutau region as building materials, showed a deterioration in their strength characteristics.

Processing of own waste for large subsoil users requires initiation of expensive, high-risk investment projects, which is associated with search for innovative technological solutions and the implementation of a full cycle of work, starting with geological exploration and ending with placing the reserves contained in the waste on state balance sheet. At the same time, profitability of such projects and the volume of mineral reserves in technogenic mineral formations are most often lower than that of projects for the development of natural deposits.

Mining companies, as a rule, do not have proven technology for processing the mining waste generated by them and dumped in dumps. The extraction of valuable components from technogenic raw materials is associated with the need to improve technological solutions and requires expensive R&D followed by the preparation of technological regulations.

Therefore, processing of technogenic objects is not priority direction of their activities and they are focused on development of explored natural deposits.

5 Problem of Involving Subsoil Waste into Economic Turnover and Ways to Solve it

5.1 Imperfection of Mining and Metallurgical Waste Assessment and Accounting Systems

Involvement of MW in economic circulation, along with preventing or reducing their formation, is an effective way to reduce emissions of pollutants and at the same time a way to rationally use the resource potential of technogenic mineral raw materials (TMRM). Its development is hampered by the imperfection of the mechanism of state regulation and administrative regulation of the sphere of handling MW. One of the manifestations of this imperfection is the current methods of assessing and accounting for MW.

The accepted system for assessing mineral resources is dominated by the geological approach, which is based on requirements for certain content and volume of useful components, but not for the technology of their extraction and processing.

Waste accounting (collection, compilation, systematization, storage, registration of information) is one of the functional components of waste management at all stages of its life cycle and is carried out for the purposes of: information exchange to ensure intra- and interdepartmental interaction between management bodies in charge of waste; providing information resources to interested parties; formation of business information environment.

Blocks to creating effective accounting system, and therefore information environment for ensuring inclusion of waste in economic circulation, are: double regulation of activities related to civil waste; lack of comparability of classification characteristics of waste as accounting objects; lack of intradepartmental interaction and procedures for transferring information about mining waste to interested parties; inconsistency of accounting systems.Dual regulation is manifested in the fact that accounting of mining waste in terms of waste from extraction and primary processing of minerals is carried out within the framework of environmental and mining legislation according to various classification criteria. The main parameters characterizing waste from the perspective of environmental legislation are the type, volume, content of harmful components, hazard class, and waste disposal sites.

Although regulatory framework of mining legislation operates with the term "waste from mining and related processing industries", it does not define it, obviously since the objects of accounting are not masses of waste, but the *mineral reserves and minerals contained in them* components whose accounting is provided State balance of reserves.

Information about the most significant technogenic mineral objects (technogenic deposits) is contained in the State Cadastre of Mineral Deposits, which is maintained in the Code of the Republic of Kazakhstan "On Subsoil" (Code 2017) to ensure the development of state and regional programs for the geological study of subsoil, the integrated use of mineral deposits, rational placement of enterprises for their production. It contains information on each deposit, including the underlying mineral resources, the components contained, as well as the mining, hydrogeological, environmental and other conditions of deposit development.

Compilation and entry of information into the deposit cadastre is carried out on the basis of data: state reports submitted by subsoil users carrying out exploration and production; conclusions of state examination of mineral reserves; territorial balances of mineral reserves received from territorial funds of geological information; geological reports on prospecting, evaluation and exploration of deposits.

Differences between accounting systems also appear in the principles of accounting for the movement of civil goods. According to environmental legislation, it is carried out according to stages of the waste life cycle (starting with its collection and ending with recycling or neutralization and burial). Mining legislation provides for different system based on accounting for movement of mineral reserves (including reserves in waste from mining and related processing industries).

Inventory accounting is carried out by placing them on the state balance sheet, transferring them from on-balance sheet to off-balance sheet (and vice versa), and writing them off from the balance sheet in the form of losses. Thus, at present, accounting for waste from mining and related processing industries and mineral reserves contained in this type of waste are independent, virtually unrelated accounting systems, which makes it difficult to exchange information and coordinate the interdepartmental interests of the two government bodies regarding waste.

Activities of subsoil users in the use (utilization) of the waste they generate and their disposal are also becoming more complicated. Access to accumulated man-made waste is limited for organizations, including small mining businesses, that either have technologies for processing such waste and are interested in applying these technologies, or have the appropriate competencies and are interested in developing innovative technologies for processing man-made waste, which are secondary mineral resources (SMR).

Effective way to solve the problem of duality and incompleteness of accounting for MW and TM, we can consider creating them on single methodological basis of the system of TMF cadastres, including regional cadastres and a specialized departmental cadastre. This will make it possible to combine disparate objects into a single system, available for study by interested users, including small mining enterprises. The purpose of the TMF inventory system is not only to perform accounting functions, but also, mainly, providing potential investors and other interested parties with sufficiently complete information about TMF potentially suitable for processing and extraction of valuable components. Duality and imperfection of systems for assessing and accounting for mining waste is closely related to the problem of double regulation of their use.

5.2 Controversies of Ownership and Use of Mining Waste

One of the main factors in the turnover of mining waste and associated processing industries is the *ownership* of this type of waste, which determines fundamental possibility of their involvement in economic circulation. The activities of subsoil users related to use of waste they generate are subject to two branches of law—mining and environmental—and are regulated simultaneously by three laws of the Republic of Kazakhstan (Code 2017; Enviromental 2007).

Subsoil legislation considers waste mining industries from standpoint of expanding mineral resource base, and waste legislation considers these same wastes from standpoint of their negative impact on the environment.

In practice, state subsoil management bodies consider *use* of mining waste as integral part of the mining process. It follows from this that the license for the right to use subsoil can only stipulate "right to use waste and condition defining this right," but not the ownership of waste, which is assigned to the state along with the right of ownership of subsoil. The Law of the Russian Federation (2007) "On Subsoil", in essence, equates mining and processing waste to immovable subsoil plots, the conditions of use of which are fully determined by the issued licenses.

In relation to mining waste (including enrichment waste) legislation on subsoil conflicts with the Civil Code of the Republic of Kazakhstan, which recognizes waste as movable property and gives ownership rights to it to the one who generated it.

Dual legal regulation of management of mining waste does not contribute to growth of processing volumes and complicates the implementation of the state's environmental policy, which, as one of the mechanisms for solving the problem of safe waste management, provides for their use by involving them in repeated economic circulation. Existing legal conflicts prevent the inclusion of a mechanism for active involvement of TMRM in the economic turnover and make such involvement economically unprofitable, as noted by many researchers of this problem. However, there are no clear points of view regarding ways to eliminate the contradictions that exist at the legislative level.

In more detail, in accordance with Art. 13 of the Code of the Republic of Kazakhstan "On Subsoil and Subsoil Use" solid waste containing useful components are classified as minerals. And according to the Tax Code, solid waste is also classified as "waste" since a fee is charged for its disposal at landfills. After all, the same object cannot be considered both as waste and as a deposit. Here TMF should be considered as secondary raw materials. And if solid waste materials are classified as secondary raw materials and property rights to them are considered, then it will be possible to process them. This problem of inconsistency of state strategies, along with problems in legislation, was solved by the President of the Republic of Kazakhstan Kassym-Jomart Tokayev, who united the ministries of ecology, geology and natural resources into one Ministry. And the Head of State set the new ministry to increase the investment attractiveness of the waste removal and processing sector.

5.3 Creation of Cadastral Maps and Databases of Technogenic Deposits

Despite the huge amount of accumulated and annually generated mining waste, there is no unified database on technogenic mineral formations that do not have the status of a deposit. The state policy of subsoil use in this regard is aimed at transferring the right to create TMF databases and develop cadastres of technogenic deposits to the regional level, i.e. Executive authorities of the constituent entities of the Republic of Kazakhstan have the right to maintain regional waste inventories.

Currently, work is underway to create regional inventories of TMF continues and receives new development in the light of the greening of mining industries based on implementation of BAT. The main goals of the cadastral assessment are: the formation of a regional database on accumulations of technogenic raw materials, preliminary ranking of solid waste materials according to the degree of their study and resource value, an initial assessment of the impact of solid waste deposits as objects of mining waste disposal on the environment.

Methodological focus of the cadastral assessment of TMF should be focused on solving the following main tasks:

1. formation of updated regional information database on man-made mineral objects by collecting and systematizing all available archival, stock, published materials and reporting information of enterprises and other organizations;
2. development of classification of technogenic deposits, drawing up maps and diagrams of their territorial location;
3. development, implementation and improvement of the existing system of passport registration of man-made objects with the compilation of a territorial cadastre of man-made deposits;
4. justification of the principles for ranking man-made objects according to their resource value and environmental hazard;
5. analysis of current state of mineral resource base of technogenic formations in the region with an assessment of the degree of their geological-technological and environmental-economic knowledge, with the identification of areas of industrial use and possible investment attractiveness.
6. development of recommendations on the methodology for additional study and subsequent systematic assessment of technogenic deposits;
7. typification of technogenic deposits with the possible identification of geological and industrial types.

The structure of the TMF inventory and the nature of the tasks solved with its help are not normatively established and are determined by many factors, such as the goals of applied research, types of extracted minerals, the degree of impact on the environment, etc. For example, one of the purposes of studying cadastral information may be to determine the elements of differentiated taxation for mineral extraction tax depending on the types of minerals in the extraction of polycomponent ores and the integrated use of mineral raw materials.

The rational use of MW as secondary mineral resources is one of the most pressing economic and environmental problems in the regions of Kazakhstan.

Changes may be made to the cadastre based on the ideas of cadastre users, ministries, departments, design organizations, and industry research institutes. Cadastral registration should be based on objective data from primary documentation that is available in organizations, and use statistical reporting data, the results of the work of design and scientific organizations, technical and economic calculations and other necessary information (Nurpeissova et al. 2023).

Rational use of mineral resources is a process of balanced economic and social changes, in which the exploitation of natural resources, including the use of MW as secondary mineral raw materials, environment preservation, industrial and environmental safety are systemically coordinated and strengthen the current and future natural resources based on attracting investments potential of the region.

One of the factors for the rational use of waste from mining, primary processing and processing of mineral raw materials is creation of *cadastre of mining waste*, which industries, descriptions of their qualitative and quantitative characteristics, location shown on maps, degree of hazard (toxicity), industrial value.

Thus, MW cadastre should be specialized database on waste from mining, ore dressing, metallurgical processing, as well as waste from supporting and auxiliary industries. This database should be created through the systematic collection, processing, storage and updating of qualitative and quantitative information on waste in the form of accounting and statistical data, cartographic materials and other documents received from mining and metallurgical enterprises by regional subsoil management authorities and systematized with the involvement of specialized scientific organizations.

As a rule, technogenic (secondary) mineral resources (MR) includes mineral waste from development of one deposit by one mine or enrichment waste from one concentration plant. At the same time, technogenic waste consists of the same types of rocks, has similar phase and mineralogical composition, physical and mechanical properties, and genesis.

For example, off-balance and associated ores, if they are stored separately, form separate TDs; dumps of the same type of overburden rocks in a mine are combined into one TD, despite the different placement of dumps on the mine territory.

Separately located tailings ponds will represent separate technogenic mineral formations. Overburden rocks from open-pit mining and tunneling rocks from underground mining, stored together at the mine, are combined into one TMF (Galperin et al. 2012). Example of separate storage of solid waste at the Donskoy MPP is shown in Fig. 4.

Database includes twenty-one interconnected structural tables that have a name, purpose, and contain information about AD fields. Monitoring and updating of the database is carried out according to certificates from mining and processing plants on the placement and accumulation of waste production in TD, from published reports and articles on the results of scientific research of individual institutes from completed reports, according to laboratory and field experiments of the institute to determine the physical, mechanical and technological properties of technogenic raw

Fig. 4 Separately stored TMF

materials, as well as final reports on research plans and contract work are received. Monitoring of replenishment of technogenic raw materials according to certificates from mining and processing plants is carried out once every 3 years.

One of the disadvantages of AD is the incomplete amount of data for some enterprises. With transition to market relations, some technical, especially economic, indicators became classified as information classified as trade secrets.

To accumulate, update and replenish information arrays about TMF is proposed to create block of mining information system, for example, of some area, which will be designed to form a set of reference, analytical, expert and other assessments when solving various mining-technological, socio-economic, managerial and environmental problems, as well as to predict the consequences testing TD, preventing occurrence and development of emergency situations, etc.

It is obvious that development of human society is impossible without interaction with environment, and, consequently, impact on nature without use of natural resources. However, the urgent need for fundamental restructuring of modern environmental management technologies has become apparent. This is due to the fact that sectoral division of responsibility for produced waste does not allow creating a sufficiently effective system for integrated management of industrial waste, ensuring maximum possible use of industrial waste as secondary raw materials (production of additional materials).

The need for development (processing) of TMF is dictated by the fact that TMF occupy vast territories, easily blown away material, and are a source of increased environmental risk for the MMC regions. The constant increase in volumes of various types of waste generated in mining and processing industries and their storage in storage facilities and experience of using such objects in industry allows us to consider them as sources for obtaining secondary raw materials and building materials.

To determine the possibility of using technogenic sources as secondary resources, Satbayev University is conducting a set of scientific works on the study and processing of solid waste (Project AR14871694 and Project BR21882292).

Job objectives:

- surveying of solid waste to create cadastral maps and determine their actual volumes;
- sampling of solid waste to study their composition with the participation of the customer;
- conducting research on solid waste in order to determine the direction of their disposal;
- improvement and implementation of technologies for the development of solid waste with complex extraction of valuable components.
- development of technology for producing effective building materials, such as: obtaining a silica component to produce *Portland cement; production of cinder blocks; production of cellular concrete; obtaining decorative and facing building materials; creation of foundations for highways.*

To carry out the work, following divisions of KazNITU named after K.I.Satpayev will be involved:

- Department of Mine Surveying and Geodesy (MD&G), equipped with a complex of modern surveying and geodetic instruments (robotic electronic total station, 3D laser scanner, drones, aerial photography from UAVs—unmanned aerial vehicles) and software) determines the occupied areas and volumes of solid waste for mapping;
- laboratory “Geomechanics and Geotechnology”, equipped with modern laboratory equipment determines the physical, mechanical and chemical properties of MW;
- laboratory of the Department of Metallurgy and Enrichment will extract useful components from MW;
- to produce building materials, KazNITU named after K.I.Satpayev is being built the plant.

6 Conclusions on the Second Chapter

1. In the modern world, rational use of subsoil resources is one of the principles of sustainable development. Requirement rationality applies equally to both natural (primary) and and to technogenic (secondary) mineral resources. Resource potential mining waste accumulated on the territory of the Republic of Kazakhstan and related industries are comparable to the resource potential of large solid mineral deposits. At the same time, mining and processing activities mineral raw materials, increased accumulation accompanying this activity technogenic waste has a serious negative impact on ecosystems of the country's mining regions.
2. The growth of waste processing volumes is hampered by dual legal regulation of the use of mining waste, which makes it difficult to actively involve TMR in economic circulation. Processing measures of MW are not effective enough; existing procedure for obtaining use rights to develop technogenic deposits and current taxation demotivate investors and interested small and medium-sized mining businesses.
3. Kazakhstan companies carry out mining as a rule, only of basic minerals. Associated minerals and components obtained during the extraction and primary processing of mineral raw materials, largely go to waste. There is no government support for subsoil users, which would stimulate the most complete extraction of all useful components from mineral raw materials. The regulatory functions of taxes aimed at involving secondary mineral raw materials in the processing require development.
4. Despite the huge amount of accumulated and annually generated mining waste in the Republic of Kazakhstan, there is no unified database on technogenic mineral formations (TMF), work on the creation of regional cadastres is carried out separately. A methodological approach to the creation of a standard regional inventory of mining waste (MW) based on Satbaey University and formation of automatic database (AD) for TMF is proposed.

References

Begentayev M (2024) Study of the influence of technological factors on the density and strength of ash-gas concrete. NEWS Nat Acad Sci Republic of Kazakhstan Ser Geol Tech Sci (1):45–57. https://doi.org/10.32014/2023.2518-170X.292. (In English)

Bekbasarov SS (2008) Problems of rational use of natural resources. Poisk—Almaty (3):135–139. (In Russian)

Bykhovsky LZ, Sporykhina LV (2011) Technogenic waste as a reserve for replenishing the mineral resource base: state and problems of development. Min Res Russia Econ Manag (4):9–15. (In Russian)

Code on subsoil and subsoil use Republic of Kazakhstan dated 27.12.2017. №. 125-VI RK (In Russian)

Satpayev KI (1962) Research of the Academy of Sciences of Kazakhstan on the integrated use of non-ferrous metal ores. Bull Acad Sci KazSSR (12):3–11. (In Russian)

Estemesov ZA, Bekbasarov SS (2004) Classification of waste. Min Bull Uzbekistan. Novai (4):60–63. (In Russian)

Estemesov ZA, Bekbasarov SS (2009) Waste and its use in the production of building materials. CLTBM, Almaty. 190 p. (In Russian)

Environmental Code of the Republic of Kazakhstan (2007) Astana, (No. 212-SH). (In Russian)
Galperin AM, Kutepov YI, Kirichenko YV (2012) Development of technogenic massifs at mining enterprises: monograph. M.: Mining Book. 336 p. (In Russian)
Law of the Republic of Kazakhstan "On Environmental Protection" dated 05.07.2007 (In Russian)
Nakipelov AD (1999) Forward from liberalism. Newspaper "Kommersant", No. 24.104. (In Russian)
Nurpeissova M, Z.Estemesov Z, Lozinsky V, Ashimova A (2023) Industrial waste recycling—one of the key directions of business development. NEWS Nat Acad Sci Republic of Kazakhstan Ser Geol Tech Sci 2(458):193–205. https://doi.org/10.32014/2023.2518-170X.309. (In English)
Pevzner ME (2008) Ecology of mining production. M.: Nedra. 235 p. (In Russian)
Project AR14871694 (2022–2024) Development of technology for processing ash and slag waste from a thermal power plant to produce popular building materials. (In Russian)
Project BR21882292, Integrated development of sustainable construction industries: innovative technologies, optimization of production, effective use of resources and creation of technological park). (In Russian)
Reimers NF (1994) Ecology. M.: Young Russia. 387 p (In Russian)
Saparova M, Ryskulbek A, Koyshybaeva M (2016) Analysis of the municipal solid waste management system of the city of Almaty. Collection of articles of the VII International Scientific and Practical Conference, resp. ed. N. F. Kuznetsova. 65–67 pp. ELIBRARY ID: 27547873. (In Russian)
Solid waste (1979) Edited by Ch. Mantell: trans. from English E.G. Teterina, A.S. Skotnikova M.: Stroyizdat. 519 p. (In Russian)
Sultanbekov TK (2001) Ecological and technological basis for the production of dry construction mixtures using natural and technogenic raw materials. Abstract of doctor of technical sciences: 25.00.36, Almaty, 33 p. (In Russian)

Experience in Using MMC Waste for Production of Construction Materials

Abstract Comprehensive review of both international and domestic research on the utilization of various types of industrial waste in the production of building materials is presented, along with the results of studies conducted by the Mining and Metallurgical Institute of Satbayev University. In particular, tailings from the beneficiation plant at the Akzhal mine were investigated as a filler component in mixtures used to reinforce quarry slopes weakened by fractured rock, as well as compromised inter-chamber pillars and ceilings in underground workings.

Reinforcement of the weakened rock mass is achieved by injecting specialized compositions into fractures, which, upon setting and hardening, significantly improve the shear strength and stability of the surrounding rock.

Research established that the developed backfill mixtures, formulated with high adhesive strength using beneficiation tailings, support the implementation of waste-free technologies in the processing of mineral raw materials. This approach not only reduces the production costs of both industrial outputs and construction materials but also contributes to environmental sustainability.

As a result, efficient solutions have been proposed for the stabilization and consolidation of fractured rock in both open-pit and underground mining environments, based on the reuse of industrial waste. The significance of these findings lies in expanding and replenishing the raw material base of the construction industry through the recycling of man-made waste and the promotion of resource-efficient technologies.

These results have practical implications for enhancing industrial safety in mining operations and minimizing the environmental risks associated with subsoil development. The large-scale integration of industrial waste—particularly from mining, metallurgical, and energy sectors- into the production of construction materials proves to be both economically beneficial and environmentally responsible, addressing pressing ecological and social issues.

M. Nurpeissova et al., *Development of Technology for Processing of MMC Waste*,
SpringerBriefs in Earth Sciences, https://doi.org/10.1007/978-3-032-12976-5_3

1 Analysis of the Current State and Prospects for Recycling Technogenic Waste for the Production of Building Materials

Waste from mining enterprises is the remains of raw materials, materials and semi-finished products formed during the manufacture of products and wholly or partially lost their consumer properties, as well as products of physical and chemical processing of raw materials, production of which was not purpose of production process and which can later be used as finished products after appropriate processing or as raw material for processing.

In many coal and ore deposits, overburden rocks from the upper horizons are represented by loose sediments, which include various clays, sands, chalk and marl rocks. Volume of overburden formation in open-pit mines ranges from 0.3–0.5 t/t of ore for rich iron ore deposits of Australia and South America to 20 t/t of ore for rich deposits of non-ferrous metals.

Clayey raw materials of overburden rocks are often represented by clays of the Devonian and Jurassic periods and Quaternary loams. These materials can be used to make cement, red building bricks, expanded clay, drainage pipes, mineral wool and mineral pigments, as well as agloporite.

Sandy deposits, as a rule, are confined to the Jurassic, Carboniferous, Devonian and Cretaceous systems. Sands can be used to produce silicate wall materials, mortars, railway and road ballasts, molding sands, etc.

Carbonate rocks composed of various limestones and dolomites are suitable for producing dry mineral paints, cement, lime, and various types of chalk. In turn, technical chalk is a raw material to produce many commercial products: paints, rubber, electrical products, paper, soda, mineral feed and fertilizers.

The most important resource for fireproofing and construction industries is marl, a dense material that is greenish-gray when wet and light gray when dry. Marls are a peculiar conglomerate of fragments of fauna, powdered calcium, quartz and clayey hydromicas. Prospects for using marls as binding components in the preparation of iron ore materials for blast furnace processing.

During open-pit mining of ore deposits, mining often also involves accompanying nonmetallic rocks, which are a potential resource for building materials—crushed stone and sand.

In general, level of overburden use throughout remains very low and, as a rule, does not exceed 3–6% of total production.

Thus, mining waste is mainly overburden rocks placed in dumps or waste heaps, and sludge from processing plants.

Industrial waste can participate in the life cycle of manufactured products in different ways depending on type of resource cycle in which they are included. The life cycle of a product is understood as period that begins with justification of research related to creation of product and ends with its transfer to consumer waste as a result of its loss of consumer properties due to physical or moral wear and tear, as well as after a certain shelf life. Product life cycle is divided into four main stages:

identification of a natural resource, production of products, operation of products or their consumption, handling of secondary material resources, i.e. production and consumption waste (Nurpeisova et al. 2023a).

Technogenic waste is industrial waste, which includes solid, liquid and gaseous waste generated at enterprises in process of obtaining the final product from raw materials. First of all, man-made waste is the remains of raw materials, materials and semi-finished products generated during production of products or work and which have lost completely or partially their original consumer properties.

The total amount of waste accumulated to date and the mass of waste generated annually is one or two orders of magnitude higher than the total capacity of national economic enterprises.

Accumulation of technogenic waste, on the one hand, leads to environmental pollution, and on the other, is consequence of irrational use of primary natural resources, majority of which are limited and exhaustible. Recycling industrial waste can become a source of large quantities of raw materials for the chemical, construction and metallurgical industries.

In this regard, utilization of technogenic waste is urgent environmental and economic task. Main industries in which technogenic waste can be used as secondary resources include production of building materials (cement, bricks, blocks), metal production, agro-industrial complex, glass, chemical and petrochemical industries. Currently, use of technogenic waste in construction is the main direction of processing technogenic waste, which allows processing of various types of waste in large quantities to produce products of sufficiently high quality.

Purpose of these studies is to analyze current state and prospects for development of various areas of utilization of industrial waste for production of building materials.

1.1 Utilization of Mining Waste

Dumps formed after mining contain large amounts of raw materials valuable for construction industry, such as chalk, clay, sand and others. Dumps from ore processing enterprises contain even more valuable components; non-ferrous metal ores, for example, are used in production of composite binders, refractories, facing materials, mineral fiber and other types of products (Galperin et al. 2012).

Now, waste dumps from processing plants are practically not used. However, recent research has discovered possibility of using waste from processing and mining in the production of porous aggregates for ceramic products, sand-lime bricks, plaster and masonry mortars, which are actively used in construction. This use of enrichment waste allows not only to reduce the cost of production and create new materials with unique properties, but also to significantly reduce environmental pollution.

Some slag melts, as a result of silicate decomposition, turn into a fine powder, "blast-furnace flour," which almost entirely consists of hydraulically active belite (dicalcium silicate phase C2S) and can be used as active mineral additive without additional grinding, which is very economically feasible (Laskorin et al. 1984).

1.2 Utilization of Ash and Slag Waste

Value of ash and slag waste is high enough to simply throw it away. The most important for construction industry and the first place in terms of volume among ferrous metallurgy wastes are blast furnace slags—by-product of smelting of cast iron from iron ores—blast furnace, open hearth, ferromanganese. Their composition can include up to 30 different chemical elements, mainly in the form of oxides, among which the largest quantities contain SiO_2, Al_2O3, CaO, MgO, and FeO, MnO, P_2O_5, V_2O_5, etc. are present in smaller quantities. The first mastered method of blast furnace processing slag is granulation. Essence of this process is the rapid cooling of the molten slag, due to which slag acquires a glassy structure and high chemical activity.

Ash and slag waste from thermal power plants is most widely used in the cement industry as a slow-hardening independent binder for road construction, as an active hydraulic additive in combination with inorganic, bitumen or polymer-bitumen binders, as a filler or low-active additive to replace part of the cement. Active use of ash and slag waste in production of sand-lime bricks can significantly reduce the use of lime (savings can range from 10 to 50% depending on composition) and sand (savings of about 30%). Bricks made using this technology are lighter than conventional ones by an average of 10%. In the production of ceramics, fuel ash and slag are often used as burn-out and waste additives, and in the production of ash ceramic products they are the main component. Main advantages of products made from ash ceramics are low abrasion, as well as high resistance to heat and acids. The largest amount of ash and slag waste is used in road construction, where it is used to fill uneven terrain, and also as a filler for asphalt concrete pavements. Ash is also used as a filler in the production of mastic for rolled roofing materials (Litvinov et al. 1993; Low-waste 1990).

1.3 Utilization of Sludge Waste

Sludge waste includes sediment from suspensions obtained in metallurgical and chemical industries as a result of processes carried out by hydrochemical methods. This type of technogenic waste is a source of large reserves of secondary raw materials to produce building materials.

Main type of slurry waste is drilling cuttings, which is understood as an aqueous suspension in which solid particles are substances formed during the destruction of the surface of broken rock and during abrasion of tools lowered into wells. The solid component of drill cuttings can also be represented by clay minerals.

Now, technologies have been developed for recycling waste from galvanic production to produce filler for various building materials and products, for example: ceramic tiles, expanded clay, silicate and ceramic bricks, synthetic rubber, concrete, expanded clay, etc. Utilization of galvanic waste by adding it to the asphalt concrete mixture allows approximately save 30% crushed stone filler.

1.4 Glass Waste Utilization

Glass waste (cullet) can have a variety of origins—glass bottles, broken sheet glass, fiberglass, defects in the production of mirror glass. There are many options for using them, but the most used are the following:

- Glass wool, which is an excellent heat and sound insulator. During its production, cullet is melted into a special fiber, which is the basis of the product. Waste use significantly reduces the cost of fiberglass, which contributes to the greater distribution of this universal insulation;
- Thermal insulation, which primarily includes foam glass, which is a glassy material penetrated by numerous uniform pores. It is characterized by low volumetric weight, low thermal conductivity coefficient and low water absorption. Also, based on cullet, foamed silica, foam concrete, foam glass-crystalline materials, etc. can be obtained (Snurikov 1986);

Liquid glass (silicate adhesive), which is a universal material and widely used in various industrial and household areas. In construction, liquid glass is used for waterproofing, as an additive to concrete, etc. Cullet replaces sand in the production of liquid glass, which reduces the cost of production.

- Interior tiles produced from glass waste using the following technology: glass, crushed to the required size, is mixed with a special resin. The finished composition is poured into molds and kept in air to harden. Technology is very simple and allows you to produce tiles of various sizes and shapes. Tile has "expensive" appearance and has water-repellent properties.

Along with listed areas, glass waste can be used as filler in production of facing products and as additive, which is source of glassy phase in production of self-glazing building ceramics and other types of ceramic products (Melkonyan 2017; Toropkina and Kalinkina 1983).

2 Study of Technogenic Waste to Obtain Building Materials

Growing scale of construction in Kazakhstan requires a significant amount of mineral raw materials for building materials industry. Intensification in this direction involves the use of industrial waste instead of primary natural resources to reduce the cost of building materials. Use of solid mining waste in building materials industry is more economical compared to production of building materials based on special mining of mineral raw materials. Review of existing scientific works in this area shows that there is significant world practice of conducting research on use of technogenic mineral raw materials. Thus, in foreign countries, mining waste is used to produce bricks, concrete, and glass ceramics. Similar studies on use of mining waste to obtain construction materials and products are being carried out in neighboring countries (Golik and Tsidaev 2015; Lygina et al. 2017).

Purpose of this work, carried out by Mining and Metallurgical Institute of KazNITU named after K.I.Satpayev, is to consider possibility of integrated use of industrial waste in areas where they are located (Rysbekov et al. 2021).

To determine possibility of using technogenic waste as secondary resources, it is necessary to carry out set of following works:

1. Surveying technogenic waste to map and determine its actual volumes;
2. Sampling of technogenic waste to study its composition;
3. Conducting studies of the composition of samples in order to determine the direction of their utilization;
4. Improving waste processing technologies to extract valuable components from them.
5. Development of technology for producing effective building materials from waste rock.
6. Production of pilot industrial batches of building materials.

Study of composition of technogenic waste was carried out using modern methods of physical and chemical research: X-ray, differential, thermal, petrographic, chemical and appropriate equipment.

Constant increase in volumes of various types of waste generated in the mining and processing industries and their storage in storage facilities and the experience of using such objects in industry allows us to consider them as sources for obtaining secondary raw materials and building materials.

One of the enterprises where non-metallic rocks and tailings are formed and where research is currently being carried out is Akshatau Mining and Processing Plant JSC, which receives feedstock from the Akzhal mine.

Results of survey of workings at the Akzhal mine (in the quarry and underground horizons) showed that the largest number of fallouts are confined to fractured rocks, and the volumes of fallouts increase as the workings stand. Observations of workings driven through fractured rocks revealed that they were stable for a month. After 2–3 months, cracks up to 10–15 cm in size are formed. Spalling and fallouts develop within 6 months, roof collapse occurs in the form of domes. This dramatically increases the volume and labor intensity of tunneling work, as well as the costs of securing and repairing workings.

To prevent collapse of workings driven through fractured rocks, anchor supports with a metal mesh of chain-link and shot concrete are used. However, delamination of roof rocks of transport roadway and significant destruction of rocks indicate that this support does not solve problem of ensuring stability of workings and does not prevent development of deformations.

As a result, after 2–3 years of standing workings, the lining is destroyed and major repairs are required. Therefore, effective solution to issue of fastening and controlling the geomechanical properties of rocks is of particular importance for the edge massifs of a quarry and for underground workings driven through fractured rocks. One of the most common methods of hardening is rock cementation. Cementation of rocks in quarries begins from the upper platform of the ledge, where fans of vertical and inclined wells are drilled. Cement mortar is injected into them until the mass is completely saturated.

In this regard, main characteristics of waste from processing plant of the Akzhal deposit were studied, the X-ray pattern and diffraction characteristics of which are shown in Fig. 1, from which it is clear that they consist of calcite, therefore, the X-ray pattern shows reflections (peaks) characteristic of $CaCO_3$, with interplanar distances

d/n, Å: 3,8665; 3,3498; 3,0404; 2,8446; 2,496; 2,2847; 2,0952; 1,9127; 1,77; 1,6287; 1,60; 1,5236; 1,4393.

Chemical analysis showed that waste rock mainly consists of, %: CaO—54,6; CO_2—39,4; SO_3—2,0; MgO—1,5; SiO_2—2,5%; Fe[S_2] about 0,18. Based on the results obtained, it can be stated that the nonmetallic rock of the Akzhal deposit consists of limestone (CaCO3)—about 95…97% and silica (SiO2)—about 2.5…3%.

Derivatogram (Fig. 2) shows only one endoeffect at 950 °C, representing the decomposition of $CaCO_3$ into CaO and CO_2 according to the equation: $CaCO_3 = CaO + CO_2$. Weight loss reaches more than 305% of the sample weight.

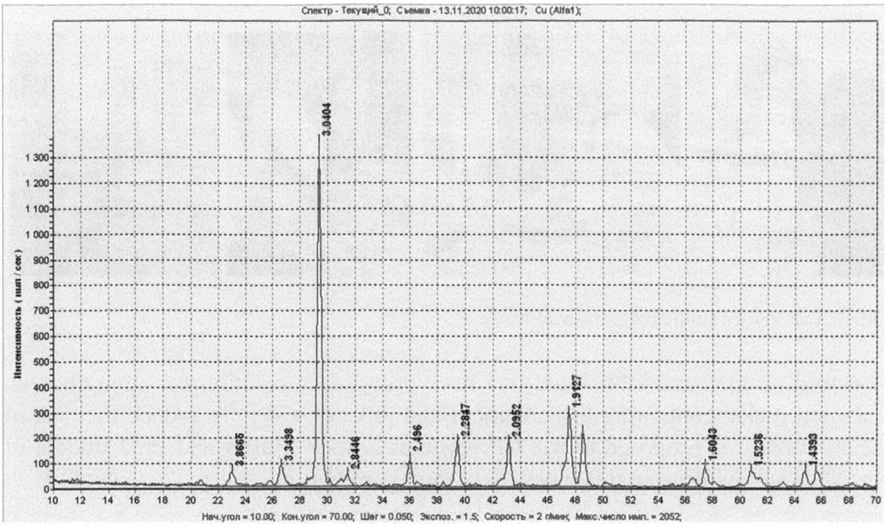

Fig. 1 X-ray diffraction pattern of Akzhal mine tailings

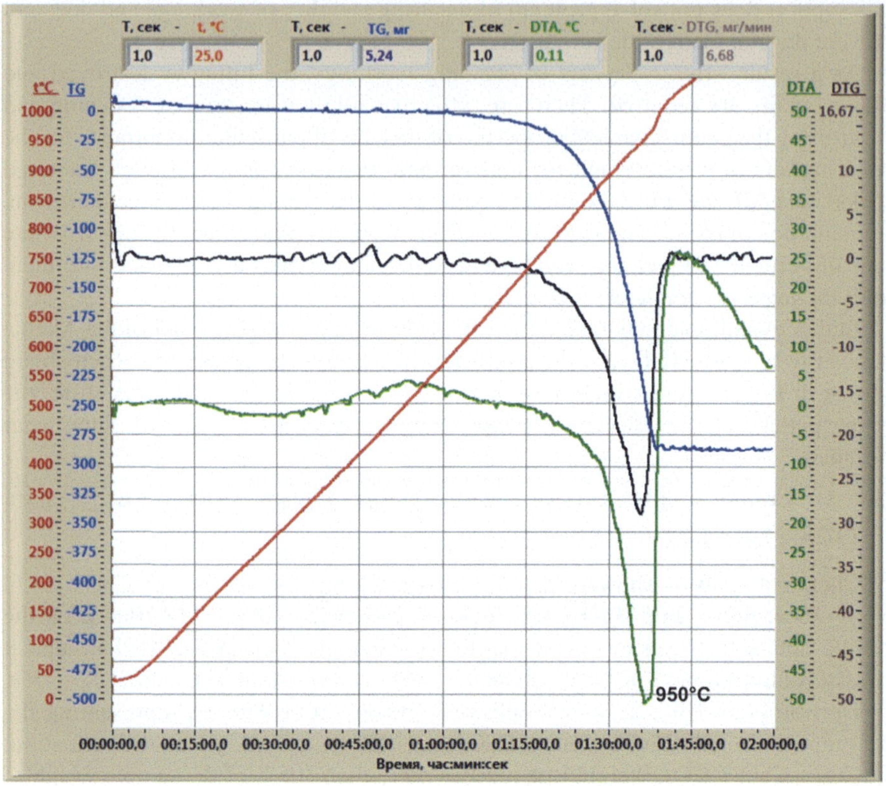

Fig. 2 Derivatogram of Akzhal mine tailings

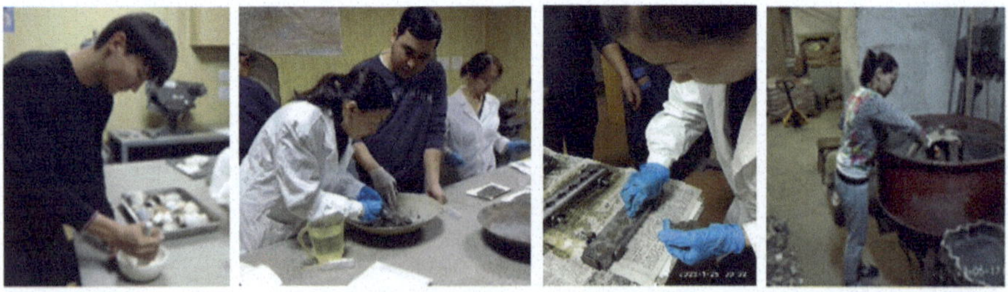

Fig. 3 Laboratory studies of industrial waste

Based on the results obtained, we have proposed a solution for strengthening fractured rocks, containing filler, cement and process water. To reduce the cost of the solution, it is proposed to use CP waste as a filler. Master and PhD student of KazNTU are involved in this work. K.I. Satpayev (Fig. 3).

Fig. 4 Construction materials produced from MMC waste

Use of waste from mining and metallurgical complexes is one of the pressing problems not only in Kazakhstan, but in all countries with developed economies. Recycling of industrial waste can become a promising area of business development in Kazakhstan for the production of competitive building materials and make an important contribution to improving the environmental situation (Fig. 4).

Use of secondary raw materials will reduce need for primary mineral resources, and there will be no need for specialized quarries for development of clays, sands, etc., which disturb the natural landscape.

Development of technologies for production of building materials based on technogenic waste, contributing to development of industrial and innovative potential of state, careful attitude towards natural resources and environment, should be considered as the most important scientific and practical task, the solution of which is directly related to environmental safety when disposing of billions of tons of waste in industrial regions. Economic efficiency of waste use is determined by the fact that technogenic raw materials have already been extracted from the subsoil, crushed to a finely dispersed state and placed in dumps.

3 Assessment of Suitability of Enrichment Waste from the Akzhal Quarry for Obtaining Construction Materials

Industrial processing of technogenic raw materials (waste from enrichment and processing, overburden and host rocks), close in composition to natural ones and used in traditional ways, is practically no different from industrial processing of natural mineral raw materials. Creation of effective technologies for processing technogenic raw materials, making it possible to obtain competitive products from them for various fields of industry, is urgent task.

This paragraph discusses development of mortar compositions for strengthening and strengthening quarry slopes. Strengthening rock mass in weakened areas is achieved by introducing substances into the cracks of rock mass, which, after strengthening and setting of rock, significantly increase its resistance to shear. Reinforcing substance is introduced into bulk mass under pressure, and as hardening material we consider cement resins, silicates and polymer resins (Zhalgassuly et al. 2020; Suleimenov 1996; Turkebaev and Sadykov 1998).

The most common among the methods of hardening was cementation of rocks, during working in aquifer rocks, strengthening unstable and disturbed massifs. With the help of strengthening, it is possible to slow down processes of weathering and shedding of rocks, prevent collapse of ledges and shedding of rocks from surface of slopes.

Possibility of obtaining solutions for strengthening fractured rocks and building structures has been studied. Study results of waste from the Akshatau Mining and Processing Plant are presented and possibility of using them for solutions to strengthen fractured rocks (Nurpeisova et al. 2022).

Injection of strengthening substance into tmass is carried out under pressure, and we consider cement mortars, silicates and polymer resins as the strengthening material. The most widespread among hardening methods is cementation when carrying out mine workings (underground structures) in fractured rocks. Significance of results obtained for construction industry lies in expansion and reproduction of the raw material base of building materials industry through use of mining and metallurgy waste (concentration waste) and development of resource-saving technologies.

Purpose of Work Assessment of possibility of using tailings and crushed stone from non-metallic rock of the Akzhal mine to obtain construction materials.

Object of the study is tailings and crushed stone from technogenic materials, and subject of study is the features of hardening of silicate systems when their fillers are carbonates.

Methodology Modern methods of physical and chemical research—X-ray, differential, thermal, petrographic, chemical, as well as the use of standard methods of physical and mechanical tests.

Over many years, the mining and metallurgical complex (MMC) of the Republic of Kazakhstan has accumulated large volumes of overburden waste, tailings, and slag. Millions of tons of harmful substances are released into the atmosphere and hundreds of millions of cubic meters of polluted wastewater are discharged into water basins. All this leads to serious economic, social and environmental problems.

According to modern estimates, enterprises of mining complex of Kazakhstan have accumulated over 50 billion tons of industrial waste and occupy vast territories (more than 150 km^2 of area). Every year the amount of industrial waste increases by approximately 1,5 billion tons, and at the same time, the level of use of TMF is currently low.

Expansion of the mineral resource base of building materials industry can be ensured not only by searching for new deposits of non-metallic minerals, but also as a result of the involvement of non-man-made waste of non-metallic raw materials in production. Technogenic raw materials, as a rule, require industrial processing and evaluation using effective methods and technologies that ensure their full use with maximum preservation of the environment.

Industrial processing of technogenic raw materials (processing waste, overburden and host rocks), close in composition to natural ones and used in traditional ways, is practically no different from the industrial processing of natural mineral raw materials. Therefore, use of mining waste to obtain construction materials from them is certainly an urgent task and priority. A large amount of research is being carried out in this direction by the staff of KazNITU named after K.I.Satpayev (Kuldeev et al. 2023; Nurpeisova et al. 2023b).

Comparative Analysis Growing scale of construction in Kazakhstan requires a significant amount of mineral raw materials for the building materials industry. Intensification in this direction involves use of industrial waste instead of primary natural resources to reduce the cost of building materials. Use of solid mining waste in building materials industry is more economical compared to production of building materials based on special mining of mineral raw materials.

Review of existing scientific works in this area shows that there is significant world practice of conducting research on the use of technogenic mineral raw materials. Thus, in non-CIS countries, mining waste is used to produce brick, concrete, and glass ceramics (Melkonyan 2017; Makarov et al. 2016).

Similar studies on use of mining waste to obtain construction materials and products are being carried out in the Republic of Kazakhstan (Baidzhanov et al. 2017). This work examines use of copper ore enrichment waste from Kazakhmys Corporation LLP to produce glass-crystalline foam materials with high performance properties. Scientists at the Central Laboratory for Certification of Building Materials (CLCBM) have developed a new environmentally friendly technology for neutralizing granulated phosphorus slag from hazardous gases to produce competitive binders and construction products (Estemesov et al. 2020). Use of additives from polymetallic ore tailings increases properties of cement mortars for strengthening fractured rocks. It has been established that developed embedded mixtures with a given high adhesive strength based on enrichment tailings contribute to the creation of waste-free technologies for processing mineral raw materials, reducing cost of both the main industrial products and production of building materials, and solving environmental problem of the environment.

Thus, it can be argued that waste recycling is not only necessary condition for environmental protection, but also a means of global resource and energy conservation.

Rational organization of waste recycling process in combination with efficient modern equipment makes it possible to obtain products from secondary raw materials at a cost 2–2,5 times lower than for similar products from primary raw materials, with comparable product quality.

The need to involve enrichment tailings in production is dictated by the following circumstances:

- service life of tailings dumps is limited; filling of many has already been completed or will end in the coming years;
- tailings occupy vast territories and, since they are finely dispersed and easily blown away material, are a source of increased environmental risk for regions where mining and processing complexes operate.

Since enrichment waste is a finely ground product that does not require additional grinding before use, this reduces economic costs. In addition, process of ore enrichment ensures t homogeneity of material both in chemical and mineralogical composition.

One of the enterprises where non-metallic rocks, tailings and waste water are formed is Akshatau Mining and Processing Plant JSC, which receives feedstock from the Akzhal mine.

The current and priority direction for utilization above-mentioned waste is, of course, using it to obtain construction materials.

Studies of the composition of waste from the processing plant of the Akzhal mine were carried out using modern methods of physical and chemical research—X-ray, differential, thermal, petrographic, chemical and appropriate equipment. To study phase composition of waste, methods of X-ray phase and differential thermal and chemical analyzes were used on the latest installations in leading countries (Russia, Germany, Japan, Switzerland.

X-ray phase analysis (XPA) was carried out on a DRON-3M X-ray installation (RF) and X-ray structural analysis was carried out with a JCXA-733 "Superprobe" microanalyzer (Japan) with software, scientific research to study the structure of waste—with laboratory polarizing microscope Leica ICH DM2500 (Switzerland), equipped with a powerful 100 W illuminator, which allows you to comfortably work with differential interference contrast.

Particle size analysis was performed using a Helos-KR diffraction laser particle size analyzer with Quixel attachments; dispersion analysis in an apparatus for dispersion analysis of powders ADAP type.

Derivatogram (Fig. 2) shows only one endoeffect at 950 °C, representing the decomposition of $CaCO_3$ into CaO and CO_2 according to the equation: $CaCO_3 = CaO + CO_2$. Weight loss reaches more than 305% of the sample weight.

1. Analysis of the mineral and chemical composition of the nonmetallic rock of the Akzhal lead-zinc deposit shows that it mainly consists of calcite—$CaCO_3$ (about 95...97%) and silica—SiO_2 (about 2.5%); among them there are also impurity elements of magnesium, iron, aluminum, zinc, lead, barium, etc., which are not of industrial interest, since their total content does not exceed 1%.
2. Tailings of the Akzhal processing plant ore mainly consist of calcite and silica, the oxide chemical composition of which is represented by the following individuals: CaO—54.3; CO_2–40.5; SO_3–2.3; SiO_2–1,5; MgO – 1,4 and Fe[S2]—0.13%.
3. Mine and waste process water, respectively, have the following characteristics: alkalinity – 0,45 and 0,8; hardness—11 and 12; pH -7.5 and 8.3. Moreover, the mine one is transparent, and the technological one is turbid, which contains ore enrichment tailings, consisting mainly of calcite $CaCO_3$.

Scope of application of enrichment tailings. The most promising areas for the use of enrichment tailings are:

- obtaining plaster and mortgage mixtures on their basis without additional processing;
- obtaining lime from them after firing at 1000 °C, used in various branches of the construction industry;
- obtaining filler for asphalt-concrete-bitumen mixture.

The composition of plastering and embedding mortars based on enrichment tailings is shown by:

- that solutions M 25...M 200 (~B2...B15) are light (1371...1449 kg/m^3);
- that with increasing grade (class/mortar), content of cement and water in it increases, and the consumption of tailings decreases.

It should be noted that by introducing chemical additives, including superplasticizers, it is possible to significantly improve the physical, mechanical and technological properties of solutions, which requires additional research.

Scope of application of crushed stone from non-metallic rock. From nonmetallic rock you can get:

- crushed stone used as coarse aggregate for concrete grades up to 300;
- crushed stone used as an underlying layer for the base during the construction of highways;
- filler as a mineral additive in the preparation of asphalt-bitumen mixture.

Scope of application of mine water and waste process water released during ore dressing

Suitability of mine and process water was checked in accordance with technical requirements of GOST 31108-2016. In this case, these waters were used as a water sealer for the cement-sand mixture. Mortar mixture mixed with ordinary water was taken as a standard.

From the results obtained it is clear that:

- setting of cement paste in technogenic waters occurs faster (160 and 170 minutes) than in ordinary water (180 minutes), which is a positive effect;
- the water requirement of cement paste using technogenic waters is slightly higher (29,5% and 31,0%) than using ordinary water (29,0%);
- the rate of strength gain of cement stone, regardless of the type of water mixing, is identical, however, the strength of cement stone in process water is greater (44.4 MPa) than in ordinary and mine water.

From the above it follows that mine and process water can easily replace ordinary water as a mixer for mortar and concrete mixtures.

4 Method and Solution for Strengthening Side Slopes of the Akzhal Quarry

The results of examination of condition of the sides of the Akzhal quarry showed that the largest number of fallouts are confined to fractured rocks, and the volume of fallouts increases as the workings stand. Observations of workings driven through fractured rocks revealed that they were stable for a month. After 2–3 months, cracks up to 10–15 cm in size are formed. Spalling and fallouts develop within 6 months, roof collapse occurs in the form of domes. This dramatically increases the volume and labor intensity of tunneling work, as well as the costs of securing and repairing workings.

The study of rock fracturing and development of methods for controlling their properties was carried out for conditions of the Akzhal field. Analysis of mining and geological conditions of deposit and state of workings driven through disturbed rocks showed that systems of steeply dipping cracks have a great influence on the stability of workings. Deposit is represented by strong rocks and, because of folding processes, they were broken by a large number of faults of various orientations. Among these disturbances are identified: large tectonic disturbances and small cracks. Intensity of various types of cracks in the rock mass was determined by measuring them in situ using a 1 m long rod or ruler (Fig. 5).

Nature of the cracks varies from smooth, wavy to uneven, splintered. The opening width of cracks does not exceed 3 mm and averages 1,5–1,8 mm. Based on studies performed on samples of fractured rocks, it was found that the rocks have different degrees of fracturing. The unsatisfactory condition of the sides of the quarries is due to the tendency of rocks to destruction, which is shown in Fig. 6. In this regard, new approach is needed to development of methodological complexes to increase stability of quarry sides and ways to increase stability of slope of benches by strengthening and hardening them (Begentayev et al. 2024).

The different strength characteristics of rocks in the massif and their sharp decrease upon exposure and contact with air and water predetermine need for a

Fig. 5 Measuring intensity of fracturing in a rock mass

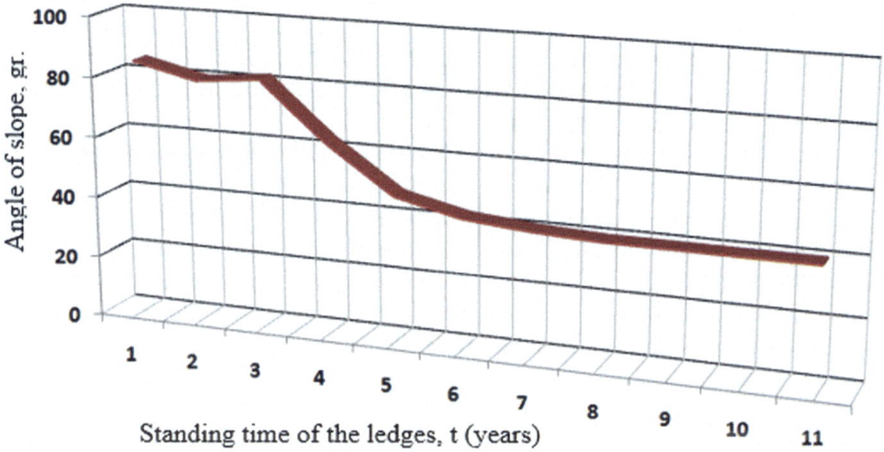

Fig. 6 Dependence of angle of inclination of ledge on time of standing of ledges

differentiated approach to managing properties of rocks with varying degrees of fracturing.

One of the most common methods of strengthening is rock cementation. Cementation of rocks in quarries begins from the upper platform of the ledge, where fans of vertical and inclined wells are drilled. Cement mortar is injected into them until the mass is completely saturated. Cement mortar is prepared based on cement and water. Technical result is to increase reliability of strengthening slope of quarry benches and to prevent shedding of rocks from surface of slope using mining and metallurgical waste.

This method of ensuring stability of slopes and benches of quarries is a complex task, solution of which should include not only determining the parameters of stable slopes, but also managing them to achieve better economic results and natural resources. Here the main task is to develop strengthening solutions at a low cost, with high strength.

In this regard, the main characteristics of the waste from the processing plant of the Akzhal deposit were studied, the X-ray diffraction pattern and diffraction characteristics of which are given in paragraph 3.2, from which it is clear that they consist of calcite, therefore, reflections (peaks) characteristic of CaCO are recorded on the X-ray diffraction pattern.

Based on the results obtained, we have proposed a method and solution for strengthening fractured rocks, containing filler, cement and process water. To reduce cost of solution, it is proposed to use tailings from enrichment factories as a filler, which are large-scale production waste and large areas are allocated for their storage (Bek et al. 2020).

Essence of this method is that screening inclined wells are drilled parallel to future surface of slope and they are blown up; create crushing zones, drill additional inclined wells, where they install reinforcement and fill them with strengthening solution. Then, wells are drilled at ledge site, anchors are secured in them, and cables are pulled between reinforcement and anchors.

During using this method in existing quarries, we were convinced that method reliably strengthens quarry bench, but is complex.

To achieve this result in proposed method of strengthening slope of quarry benches, which involves creating a crushing zone by drilling screening inclined wells parallel to the future surface of slope and blasting them; After removing blasted rock mass and cleaning surface of slope, reinforcement with hooks is hammered in so that it is adjacent to future surface of slope. Chain-link mesh is attached to hooks over entire surface of slope, and then layer of strengthening solution is applied to it.Moreover, the strengthening solution is prepared based on the tailings of quarry concentration plant (CP).

Method is carried out as follows (Fig. 7).

Based on our research, we have proposed solution for strengthening fractured rocks, containing filler, cement and process water. To reduce cost of solution, it is proposed to use tailings from enrichment factories as a filler, which are large-tonnage production waste and large areas are allocated for their storage.

Dry superplasticizing additive Neolit 400, which is produced by Neochim with high water-reducing ability and makes it possible to reduce water-binder ratio in systems by more than 20%. By decreasing water-binder ratio, durability and density of solution being developed increases, with a simultaneous decrease in shrinkage and creep deformations as the solutions gain strength. Additive is well compatible with Portland cements, cement—up to 37%, tailings from processing plants—up to 52%, superplasticizer Neolit 400 – 0,11-0,16 and the rest is water. The ratio of components shown was obtained experimentally in laboratory conditions. To determine strength, samples of 4 × 4 × 16 cm are molded from mixture and compacted on a vibrating platform for 45 s, after a day they are removed from the molds and stored in humidity conditions for 28 days (starting value), and then physical and mechanical tests are carried out, results of which are presented in Table 1.

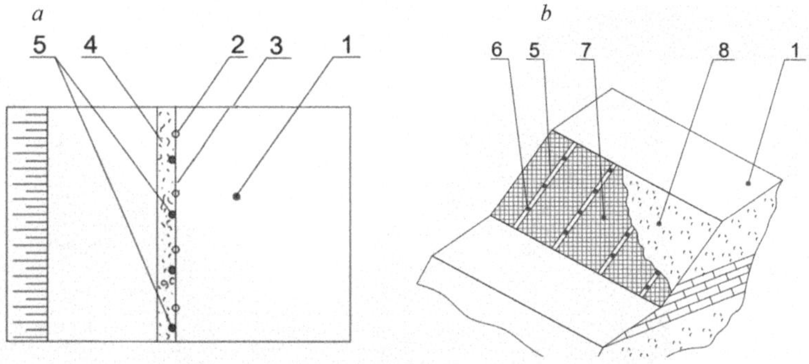

Fig. 7 (**a**) Working platform in plan; (**b**) Strengthening quarry slope in isometry

Table 1 Physical and mechanical properties of solution

Sample	Composition of solution, wt %				Tensile strength, MPa		Cone draft, mm
	cement	PP tails	Neolit 400	Water	for compression	for bending	
1	32	52	0,16	15,9	32,4	4,3	150
2	33,4	49,3	0,13	16,3	35,7	5,1	145
3	37	47	0,11	16,9	36,9	5,7	142

Thus, use of solution described above ensures strengthening of weak areas of sides and can significantly reduce harmful effects of waste from processing plants on environment. The technical novelty of the created solution was confirmed by patents of the Republic of Kazakhstan for the invention (Patent No. 36246 2023).

These studies confirmed that proposed composition of solution for strengthening fractured rock mass should be in the following ratio, wt. %: cement 32–37, tailings from processing plants 47–52, superplasticizer Neolit 400 – 0,11 – 0,16, the rest is water.

Result of strengthening the damaged area is an increase in the safety factor to a value of 1.04 (Fig. 8).

The slip line that could potentially form is also in an area of increased fracturing, but filling the cracks increases rock cohesion and stabilizes deformations.

Thus, the presence of a zone of fracturing in the near-edge massif, associated with a tectonic disturbance, causes an unstable state of the quarry side. The stability margin coefficient decreases not only below the design level (1.3), but also below the limit equilibrium level (1.0), amounting to 0.92. At the same time, the injection of a hardening mixture based on tailings from the enrichment of polymetallic ores helps to fill cracks and increase cohesion in the block structure of rocks, which increases the ability of rocks to resist shear components of the stress field.

Significance of the results obtained for the construction industry lies in the expansion and reproduction of the raw material base of the building materials industry using enrichment waste from the Akzhal deposit. Results obtained can be used

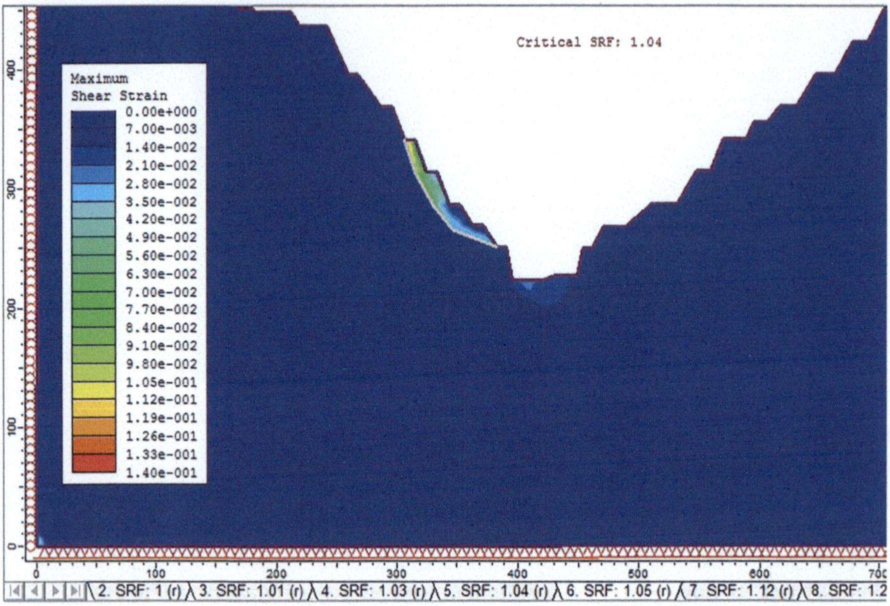

Fig. 8 Distribution of maximum shear strains and potential slip line when fixing cracks in the fault zone

to improve level of industrial safety at mines and minimize environmental risks caused by subsoil development.

5 Development of Method Producing Effective Solution Based on Enrichment Tailings for Strengthening Fractured Rock Massifs in Underground Workings of the Akzhal

Results of survey of workings at the Akzhal mine (in the quarry and underground horizons) showed that the largest number of fallouts are confined to fractured rocks, and volumes of fallouts increase as the workings stand. Observations of workings driven through fractured rocks revealed that they were stable for a month. After 2–3 months, cracks up to 10–15 cm in size are formed. Spalling and fallouts develop within 6 months, roof collapse occurs in the form of domes. This sharply increases volume and labor intensity of tunneling work, as well as the cost of securing and repairing workings.

To prevent collapse of workings driven through fractured rocks, anchor supports with a metal mesh of chain-link and shot concrete are used. However, delamination of roof rocks of transport roadway and significant destruction of the rocks indicate that this support does not solve problem of ensuring the stability of the workings and does not prevent the development of deformations.

As a result, after 2–3 years of standing workings, lining is destroyed and major repairs are required. Therefore, effective solution to issue of fastening and controlling geomechanical properties of rocks is of particular importance for edge massifs of quarry and for underground workings driven through fractured rocks.

In this regard, the main characteristics of the waste from the processing plant of the Akzhal deposit were investigated.

Analysis of the mining and geological conditions of the Akzhal deposit and the state of underground workings driven through disturbed rocks showed that rock cracking has a great influence on the stability of workings. Different strength characteristics of rocks in the massif and their sharp decrease upon exposure and contact with air and water predetermine the need for a differentiated approach to managing the properties of rocks with varying degrees of fracturing. One of the most common methods is shotcrete strengthening of disturbed areas, that is, injection of cement mortar until the mass is completely saturated.

Goal of the work. Development of effective solution based on enrichment waste to strengthen and harden disturbed rocks in underground mines by increasing their strength characteristics.

Research methodology. To determine characteristics of starting materials and composition of embedded mixtures and their physical and mechanical properties, standard methods were used, and XRF and ICS were used to identify their physical and chemical properties.

To strengthen disturbed fractured rock masses in underground mines, we have proposed an effective shotcrete solution. Technical result is recycling of mining waste and tailings from processing plants, achieving high fluidity of the solution, adhesion to rocks and strength of resulting solution.

Shotcrete reinforcement (hardening) has the following advantages over traditional methods:

- when applying mixture under pressure, particles of PP waste with small fractions are clogged into rock cracks, restoring damaged peripheral layer according to the type. As a result, additional restored rock shell is included in system of forces that counteract delamination of tmassif. This "additional" shell works in a single system with shotcrete coating, equalizing stress, increasing its stability;
- sprayed concrete coatings, applied in a small layer, reliably protect rocks from destruction, as a result of which rocks retain their properties unchanged for a long time. Under normal conditions, strength of rocks in vicinity of excavation drops over time to 60–70% of the initial value;
- movement of sprayed concrete mixture through pipes under high-speed air pressure increases dispersion of solution and increases number of hydrating grains. This helps to increase strength of shotcrete. In addition, solution activity increases due to constant compaction under influence of jet escaping from the nozzle;
- a high level of mechanization of process of strengthening disturbed surfaces ensures a 2–3-fold increase in the labor productivity of tunnelers and fasteners;

- Shotcrete coatings increase safety of mine operation by warning in advance about occurrence of danger. This is explained by fact that deformations in the shotcrete reinforcement gradually appear in form of cracks, and high adhesion to rock keeps the damaged concrete shell and pieces of separated rock from falling out. Repair, restoration and strengthening are very easily done by repeated spraying.

To achieve this result, the following raw materials were studied: Portland cement PC 400, produced by Central Asia Cement JSC (Karaganda region, Kazakhstan), enrichment tailings from Akzhalskaya OF, functional additive Reparatur, produced by Ading (North Macedonia) and polycarboxylate additive "Neolit 400" (Russia).

Justification for the choice of source materials:

1. The use of Portland cement PC 400-, produced by Central ASIA cement JSC, is due to the location of the plant for its production in the vicinity of the city of Balkhash, not far from the Akzhal mine.
2. The choice of tailings from the Akzhal enrichment plant (hereinafter referred to as APP) is explained by the following:
   - to improve ecological situation of the region's environment through their complete recycling with environmental and economic efficiency;
   - to use carbonate composition of tailings to develop special cementing solutions for purpose of more effectively strengthening rock cracks in underground workings with them.
3. Use of dispersive polymer powder (DPP)—the additive "Reparatur" from Ading (North Macedonia) in the composition of the above-mentioned solutions is associated with its properties—increasing the adhesion strength to the surface of rock cracks.
4. Use of dry superplasticizing (SP) additive "Neolit 400" (Russia) in the composition of cementitious solutions is due to its high water-reducing properties. By decreasing the water-binder ratio, durability and density of solution being developed increases, with a simultaneous decrease in shrinkage and creep deformations as the solutions gain strength.

Characteristics of Portland Cement PC 400 D0 (M400)

As can be seen from Table 2 Portland cement is represented by alite (C_3S – $3CaO{\bullet}SiO_2$, rather—$54CaO{\bullet}16SiO_2{\bullet}Al_2O_3{\bullet}MgO$), belit ($C_2S$ – $2CaO{\bullet}SiO_2$), celite ($4CaO{\bullet}Al_2O_3{\bullet}Fe_2O_3$) and felit ($3CaO{\bullet}Al_2O_3$). Alite content—about 50% by weight, C_2S – 20%, C_3A – 7%, C_4AF – about 15%, the rest is glass phase.

Chemical composition of APP waste is represented by the following oxide content, wt. % (Table 3).

Granulometric composition, physical properties and specific effective activity of natural radionuclides and enrichment tails are given in Tables 2, 3, 4 and Figure 1, from which it can be seen:

- that sand fineness modulus is 0.83, therefore, it belongs to very fine sands, which is suitable for preparing plaster mixtures;

Table 2 X-ray pattern of Portland cement

ϴ	d,A	I	Main phases	ϴ	d,A	I	Main phases
12,2	7,18	10	C_4AF	33,01		14,4	C_3S, алюмин.
14,9	5,94	0,3	C_3S	34,2		18	C_3S, C_2S
21	4,23	3,2	гипс	34,5		100	C_3S, C_2S
23,2	3,83	11,3	C_3S	37,8		11,3	C_2S
26.8	3,32	15,4	C_2S	39		17,5	C_2S
29,6	3,03	72	C_3S	41,4		45,3	C_2S, C_3S
30,2	2,94	18,5	3_2S	47,4		17,5	C_4FF
31	2,86	10,3	C_2S,гипс	51,8	1,76	51,5	C_3S
32,4	2,76	64,9	C_3S, C_2S	56,6		25,7	C_3S
33	2,73	46,4	C_3S, C_2S	60,2		20,6	C_3S

Note: ϴ - angle of reflection of the x-ray; d, A – interplanar distances in A; I-intensity on a ten point scale or intensity number (imp/sec)

Table 3 Chemical composition of APP

Oxides	Amount, wt.%
CaO	54,3
CaO_2	40,5
SO_3	2,3
SiO_2	1,5
MgO	1,4
$Fe[S_2]$	0,13

- that bulk density is 1515 kg/m^3, the true density is 2.74 g/cm^3, and the voids content reaches 44.7%;
- that the specific effective activity of radionuclides complies with the regulatory requirements of GOST 30108-94 - 120.0 Bq/kg versus up to 370 Bq/kg according to the standard.

Figure 9 shows the effect of Neolit 400 SP on strength of samples, from which it can be seen that:

- With introduction of SP, strength of samples increases by 3…27% compared to their strength without additive;
- At the same time, the more cement in the mortar mixtures, the greater the positive effect of joint venture on strength; for example, with a cement consumption of 69 kg/m^3 and with its optimal content (0.3%), the increase in the strength of samples in its presence reaches 11%, and with a cement consumption of 265 kg/m^3, the increase in the strength of samples reaches 29%;
- Positive effect of the SP content passes through a maximum, based on this, it can be stated that the optimal dosage of the plasticizer is 0,3% of the cement content.

Thus, composition has been selected to strengthen fractured rocks in underground mines. Components for preparing solution are taken in the following ratio: Portland

Table 4 Granulometric composition of APP tails

Name of indicator, unit of measurement	Actual value
Grain composition:	Complete residues on sieves, %
2,5 мм	2,5
1,25 мм	4,7
0,63 мм	7,6
0,315 мм	17,0
0,16 мм	51,4
Grain size content, %:	3,26
over 10 мм	1,96
over 5 мм	46,3
less 0,16 мм	sand class – II
Size module	0,83 sand group – thin

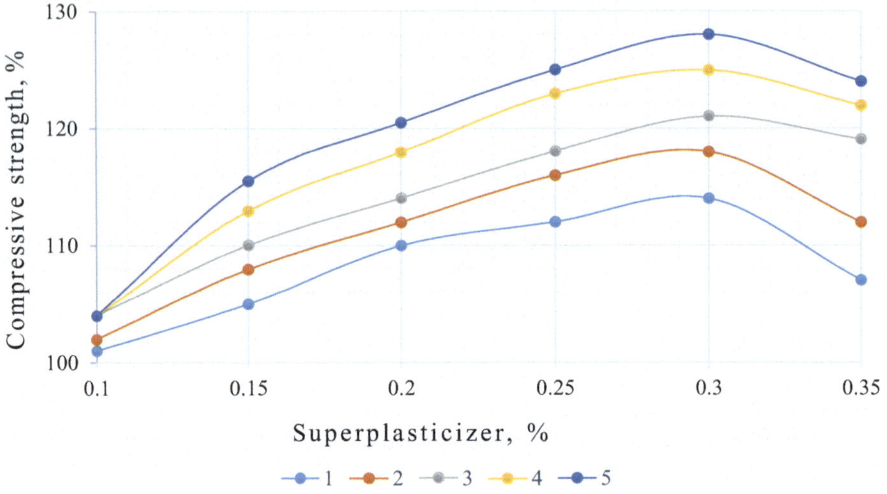

Fig. 9 Dependence of sample strength on Neolit 400 SP

cement PC 400—up to 18%, tailings from processing plants—up to 71.2%, Reparatur additive—1.5%; superplasticizer Neolit 400 – 0.3%; the rest is water (water-cement ratio is about 0.5)—9%.

After dosing components, cement, tailings and additives are loaded into a concrete mixer and mixed thoroughly. Then add water and mix. Finished composition is delivered to underground workings and injected into the cracks of disturbed underground mine workings.

To determine strength, samples of 7 × 7 × 7 cm were molded from the composition and compacted on a vibrating platform for 45 s. A day later, samples were removed from molds and stored in humid conditions for 28 days, and then physical and mechanical tests were carried out, the results of which are presented in Table 5.

Table 5 Physico-mechanical properties of the resulting solution

Solution composition, wt.%					Indicators	
Portland cement	Tails PP	"Reparatur"	Neolit400	Waste water PP	Tensile strength compression, Pa	Mixture mobility, cm
18	71,2	1,5	0,3	9	21,5	100

Technical novelty of the created solution was confirmed by patents of the Republic of Kazakhstan for the invention (Patent No. 36220 2023).

In the conditions of the Akzhal and Saryoba mines, technology for strengthening cracked sections of underground workings with sprayed concrete mortar was tested. Analysis of studies carried out in underground workings also showed that the highest technological and technical-economic indicators of fastening mine workings using shotcrete method are achieved through correctly selected composition of the concrete mixture, consider specific mining-technical, mining-geological and hydrogeological conditions of workings.

As a result of studies, it was found that tensile strength of resulting solution is 1,5–2 times higher, and adhesion to rock increases 6 times. Also, solution based on Neolith 400 has increased elasticity, crack resistance, water resistance, adhesion to rock and strength.

Originality. It has been established that developed embedded mixtures with a given high adhesive strength based on enrichment tailings contribute to creation of waste-free technologies for processing mineral raw materials, reducing cost of both the main industrial products and production of building materials, and solving the environmental problem of environment.

Scientific novelty. Positive role of calcite in obtaining effective embedded mixtures – composites – has been revealed. Mechanism of influence of tailings calcite on composition hardening is shown. Dependence of strength of samples based on PP tailings on superplasticizer was obtained. Significance of resulting effective solution for strengthening and hardening fractured rocks in underground mines lies in expansion and reproduction of raw material base of building materials industry through use of waste from the Akzhal deposit.

6 Conclusions on the Third Chapter

1. Thus, one of the effective ways to solve problem of accumulation of technogenic waste in large quantities is their secondary use as fillers or modifying additives in the production of building materials and products. As can be seen from information presented in this work, use of technogenic waste as secondary resources allows not only to reduce environmental pollution from waste, but also makes it possible to obtain a high-quality product while saving primary raw materials, which reduces production costs.

2. Processing and recycling of mining and metallurgical waste into construction materials is aimed at solving technological, environmental, social and economic problems in regions with developed mining and metallurgical industries. Rational organization of waste recycling process in combination with efficient modern equipment makes it possible to obtain products from secondary raw materials at a cost 2–2,5 times lower than for similar products from primary raw materials, with comparable product quality.
3. By crushing non-metallic rock of the Akzhal deposit, it is possible to obtain crushed stone suitable for producing concrete M 100…M 300 (class: 7.5…B22.5) and the base of highways; In addition, by firing it can be used to produce lime needed for various branches of the construction industry. In terms of its mineralogical and chemical composition, ore tailings are identical to those of nonmetallic rock; consist mainly of limestone—$CaCO_3$. Difference: nonmetallic rock is a large stone, and tailings are fine sand. Mine and waste process water are suitable for use as a water sealer in the production of mortar and concrete mixtures.
4. Solution has been developed for strengthening fractured rocks, based on MMC waste using polymer SP and DPP, which has low cost, sufficient fluidity, compressive and adhesion strength after hardening to fill small cracks and high strength. The technical novelty of created solution is confirmed by patent of the Republic of Kazakhstan for invention.

References

Baidzhanov DO, Bek AA (2020) "Green" economy in building materials. Min J Kazakhstan:45–48. http://minmag.kz/ru/2021/10/05/%e2%84%967-2020/ (in Russian)

Baidzhanov D.O., Rakhimov M.A., Rakhimov A.M. Technology for producing foam glass-crystalline heat-insulating materials based on industrial waste. Proceedings of Karaganda State Technical University, Karaganda, 2017. 73-76. https://scholar.google.com/citations?view_op=view_citation&hl=ru&user=R8JXLHwAAAAJ&citation_for_view=R8JXLHwAAAAJ:mVmsd5A6BfQC (in Russian)

Begentayev MM, Kuldayev YI, Nurlybayev RY, Nurpeisova MB (2024) Creation of solutions based on mining and metallurgical waste to strengthen cracked surfaces. In: International scientific conference «Prospects for the application of digital technologies in rational and safe subsoil use». Tashkent State Technical University, Tashkent, pp 291–296. (in Russian)

Bek AA, Donenbaeva NS, Aitkazinova SK, Nurpeisova MB (2020) Studying the strength properties of rocks at the Akzhal mine in order to strengthen weakened areas. Young Scientist J 33(323) https://moluch.ru/archive/323/73183/ (in Russian)

Estemesov ZA, Barvinov AV, Sarsenbaev BK, Tulaganov AA, Estemesov MZ, Khaidarov AM (2020) New method for disposal of granulated phosphoric slag from hazardous gases. NEWS Nat Acad Sci Republic of Kazakhstan Ser Chem Technol 3(441):6–14. https://doi.org/10.32014/2020.2518-1491.37. (in Russian)

Galperin AM, Kutepov YI, Kirichenko YV (2012) Development of technogenic massifs at mining enterprises: monograph. Mining Book. 336 p. (in Russian)

Golik VI, Tsidaev TS (2015) Methodology for using tailings from the processing of substandard mineral raw materials. Construction materials and technologies of the 21st century 12:27–29. http://www.stroymat21.ru/arch.php?file_=./2012-12/newtxt.htm (in Russian)

Kuldeev EI, Nurpeisova MB, Bek AA, Ashimova AA (2023) Prospects for technogenic waste processing for production of construction materials. Almaty: Min J Kazakhstan (4) 57–64 pp. (in Russian)

Laskorin BC, et al. (1984) Waste-free technologies for processing mineral raw materials

Litvinov VK, Dmitriev SA, Kiyarv CA et al (1993) Plasma shaft furnace for processing radioactive waste of medium and low activity. Magnitogorsk, Magnitogorsk Mining and Metallurgical Institute, NPO "Radon". (in Russian)

Low-waste and non-waste technologies (1990) Conference materials. Secretariat. (in Russian)

Lygina TZ, Luzin VP, Kornilov AV (2017) Multi-purpose use of technogenic non-metallic raw materials and obtaining new types of products from them. Fort Dialog-Iset Ekaterinburg 1:67–71. ISBN: 978-5-91128-154-0 https://www.elibrary.ru/item.asp?id=29734811&pff=1 (in Russian)

Makarov DV, Melkonyan RG, Suvorova OV, Kumarov VA (2016) Prospects for the use of industrial waste to produce building materials. Min Info Analy Bull 5:254–281. ISSN 0236-1493. (in Russian)

Melkonyan RG (2017) Environmental problems of recycling mining waste for the construction industry. Bull Sci Educ North-West of Russia 3(1) ISSN: 2413-9858 http://vestnik-nauki.ru/wp-content/uploads/2017/02/2017-No1Melkonyan.pdf (in Russian)

Nurpeisova MB, Estemesov ZA, Bekbasarov SS (2022) Processing of technogenic mineral formations is a promising niche for business. Proceedings of the international scientific-practical conference, Dedicated to the 115th anniversary of corresponding member. Academy of Sciences of the KazSSR A.Zh. Mashanov and the 100th anniversary of Academician of the Academy of Sciences of the KazSSR Zh.S.Erzhanov "Innovative technologies in geospatial digital engineering" March 18, 2022, Almaty. 191–197 pp. (in Russian)

Nurpeisova M, Estemesov Z, Lozinsky V, Ashimova A (2023a) Industrial waste recycling – one of the key directions of business development. NEWS Nat Acad Sci Republic of Kazakhstan Ser Geol Tech Sci 2(458):193–205. https://doi.org/10.32014/2023.2518-170X.309. (in English)

Nurpeisova MB, Rysbekov KB, Ashimova AA (2023b) Prospects for technogenic waste processing for production of construction materials. Горный журнал Казахстана (1) ISSN 2227-4766. (in Russian)

Patent of the Republic of Kazakhstan No. 36246 Method for strengthening a quarry slope (Bek AA, Donenbaeva NS, Nurpeisova MB, Aitkazinova SK, Nukarbekova ZM, Derbisov KN), dated 02.06.2023 г. (in Russian)

Patent for invention № 36220. Composition of a solution for strengthening disturbed rock masses in underground workings (Bek AA, Donenbaeva NS, Nurpeisova MB, Aitkazinova SK, Nukarbekova ZM, Derbisov KN) dated 05.04.2023. (in Russian)

Rysbekov КБ, Nurpeisova MB, Bek AA (2021) Use of enrichment waste to obtain building materials. In 4th international scientific and technical conference "Innovative development of resource-saving technologies and rational use of natural resources". Petrosani, Romania: University Publishing House, 16–19 pp. https://elprivod.nmu.org.ua/ua/articles/Ukraine_2021_Book_of_Abstracts.pdf. (in Russian)

Snurikov AP (1986) Integrated use of raw materials in non-ferrous metallurgy. Metallurgy

Suleimenov ST (1996) Physico-chemical processes of formation structures in building materials from mineral industrial waste. Manuscript, Moscow. 298 p. (in Russian)

Toropkina GN, Kalinkina LI (1983) Technical and environmental indicators of industrial purification of gaseous emissions of organic substances. (in Russian)

Turkebaev EA, Sadykov GK (1998) Integrated use of raw materials and industrial waste. Kazakhstan, Almaty. 140 p. (in Russian)

Zhalgassuly N, Estemesov ZA, Kogut AV (2020) Chemical and mineralogical characteristics of technogenic raw materials of mining enterprises of Kazakhstan. 20th international scientific multidisciplinary conference on earth and planetary sciences SGEM2020. Conference period: 16 August, 2020–25 August, 2020 Congress Center "Maritim Paradise Blue", Albena Resort & Spa, Bulgaria. https://doi.org/10.5593/sgem2020/5.1/s20 (in English)

Mining Waste: Raw Material Base for Obtaining Construction Materials

Abstract Currently, the volume of technogenic mineral formations (TMF) accumulated in waste dumps across Kazakhstan amounts to billions of tons. These industrial wastes exert significant environmental pressure on ecosystems, often contributing to soil, air, and water pollution. To reduce the ecological burden, it is essential to utilize beneficiation waste as a raw material for producing various types of construction materials.

This chapter presents both quantitative and qualitative characteristics of the waste generated by the Zhezkazgan and Kentau beneficiation plants and explores the potential for its practical application.

Modern construction practices are increasingly focused on introducing efficient and innovative materials that reduce labor intensity and material consumption on construction sites while enhancing the overall quality and durability of the structures. Among such materials are dry building mixtures (DBMs), which represent a product of advanced technological development.

Technological solutions are proposed for the production of DBMs and cellular concrete using ore beneficiation waste and modified diatomite. Preliminary assessments indicate that Kazakhstan's diatomite reserves exceed 200 million tons, primarily concentrated in the Mugodzhar district of the Aktobe region.

Feasibility of using activated diatomite in construction material technologies has been studied. When properly processed, high-quality diatomite can be employed as a carrier for catalysts, filtration media, and lightweight structural ceramic materials. In addition, marl has shown potential as a raw material component in the production of ceramic products for industrial use.

Review of current trends in the paints and coatings industry reveals promising opportunities for the use of water-dispersion polymer-based formulations. Modern technological advancements in this area focus on improving the performance and durability of these materials by modifying aqueous polymer dispersions with multifunctional surfactants to ensure long-lasting and effective protection.

M. Nurpeissova et al., *Development of Technology for Processing of MMC Waste*, SpringerBriefs in Earth Sciences, https://doi.org/10.1007/978-3-032-12976-5_4

1 Practice of Using Technogenic Mineral Resources

The huge scale of construction in Kazakhstan requires significant amount of mineral raw materials for building materials industry. Intensification in this direction involves use of industrial waste instead of primary natural resources to reduce the cost of construction materials. Use of solid mining waste in the building materials industry is more economical compared to the production of building materials based on special mining of mineral raw materials. Identification of raw materials suitable for use in production of scarce building materials will significantly replenish raw material base. Kazakhstan has accumulated some experience in the use of non-ferrous metallurgy waste, namely sludge in production of reinforced concrete products.

Study of experience of using mining waste showed that mainly mineral materials are used as raw materials, which can be divided into 3 main groups (Korneev and Zozulya 2010; Bazhenov and Denisov 2003; Zharko and Guzov 2011):

- siliceous materials, content more than 50% SiO_2;
- lime materials, content more than 50% CaO;
- calcareous-siliceous materials containing from 20 to 50% CaO and 20–50% SiO_2.

Scope of use of these wastes is largely determined by their mineralogical composition.

Composition of mining waste determines the ways of its expedient use and creates the prerequisites for development of the raw material base of non-metallic materials.

Mineral resource base of the building materials industry includes 500 types of non-metallic raw materials. Every year there is an increase in the demand for construction materials, which averages 10% (Goncharov and Chernyshov 2013). However, needs of construction industry are not fully satisfied in terms of volume, range and quality of products. Use of industrial waste, many of which in their composition meet the requirements for building materials, makes it possible to cover up to 40% of the need for raw materials. In addition, use of waste from various industries, including mining and processing, will reduce the cost of building materials by 10–30% compared to their production from natural raw materials by reducing the cost of energy, labor, equipment and transport. Specific weight of transportation costs in selling price of products during delivered over long distances reaches 55–65%.

Analysis of production in recent years shows that there has been a steady increase in production for almost all types of building materials. Annual increase in the production of building materials in the Republic is 24.4%. Cement production has been growing steadily in recent years. Total demand for cement products for period 2015–2020 increased from to 10,5 million tons. According to calculations, for the period until 2024.

High potential of building materials market in the Republic of Kazakhstan will remain. DBM is currently in high demand on the building materials market. Dry mixtures, unlike traditionally prepared mortars and concretes, are delivered to construction sites in dry form and mixed with water immediately before use (Sartbaev et al. 2017).

Current emergence of new modifying additives and technologies makes it possible to widely use dry mixtures in the development of new technologies for construction work. These materials are produced based on binders with modifying additives, and in some cases with granular and reinforcing fillers and pigments. Construction sands, marble chips, ground limestone, mineral or organic fibers, and metal fiber can be used as filler. Properties of resulting mixture depend on modifying additives, number of which in mixture can reach ten or more, some occupying an insignificant percentage of volume.

Currently, one of the main factors in the formation of a competitive industry is innovation, the introduction of new scientific developments into production. Experience of developed countries shows that one of effective measures for development of industry is the attraction of scientific and technical achievements into industrial production.

Technogenic mineral formations of various origins differ in physical and mechanical properties, mineralogical and chemical composition, size, density and strength. Presence of many TMF varied in composition and technical characteristics, determines the need for their comprehensive study to select sound technical and economic solutions, consider environmental factors.

The largest reserves of solid waste are concentrated in tailings dumps. Concentration waste is a finely ground product that does not require additional grinding before use, which reduces economic costs. In addition, process of ore enrichment ensures homogeneity of material both in chemical and mineralogical composition.

In this regard, we have chosen direction of research to develop technology for obtaining building materials from tailings of copper, manganese, lead-zinc, chrysotiles-bestos and chromite ores. For this purpose, samples were taken from the tailings and their chemical, mineralogical, and granulometric composition was determined.

Analysis of research results showed that the most promising for conducting research on development of competitive technology for obtaining building materials are the current tailings of enrichment of copper ores of the Zhezkazgan deposit and polymetallic ores of Achpolimetal JSC.

2 Compositions Development of Dry Construction Mixtures with Use of Ore Enrichment Waste from the Zhezkazgan and Kentau Concentration Plants

Study of the kinetics of hardening and strength of cement-sand mortars carried out by the authors on waste from the enrichment of non-ferrous metal ores from the Zhezkazgan and Achisai deposits showed that the strength of the hardening material depends on its mineral composition and is associated with the processes of interaction of finely dispersed part of waste with clinker minerals. In addition, these indicators depend onthe adsorption and adhesion activity of materials and wettability. The interaction of beneficiation tailings grains with cement mortar is determined by the structure of the crystals, the microrelief of their surface, porosity and hydrophilicity or hydrophobicity of the grains. Smooth and hydrophobic surfaces are characterized by a sharp drop in the adhesion of the binder stone to the grains of aggregate and micro-aggregate. On rough and hydrophilic surfaces, epitaxial fusion of new formations of cement stone with the contact layer of filler appears. Strength of mixture formed during hardening process also depends on strength of filler used and adhesion of binder stone to filler grains.

Results of study of granulometric and mineralogical composition of processing tailings of the Zhezkazgan concentrator plant and processing tailings of polymetallic ores of the Mirgalimsay deposit showed possibility of using this technogenic raw material as a starting material for obtaining charge of various compositions and purposes, since it is grains of carbonate minerals (calcite and dolomite) and surface of pure quartz and feldspar provide high-strength bonds with the clinker.

The main processes of properties formation of mortars are determined by interactions in system "mineral binder—aggregate—water (Gorshkov et al. 1991; Zhaparov et al. 2013), but introduction of inorganic and organic modifying additives into such system allows you to change practically all the characteristics of material and obtain mortars with specified properties, intended for use in various, including extreme, conditions. Use of modifying additives in the compositions of dry construction mortar mixtures makes it possible to change their technological properties within a wide range.

To develop technique for obtaining compositions of dry mixtures based on copper ore enrichment waste from the Zhezkazgan processing plant (ZhezPP), polymetallic carbonate ores from the Mirgalim deposit and asbestos ore enrichment waste from the Zhitikarinskoye deposit, following materials were used: Portland cement PC 400 D 20 JSC "Central AZIA Cement" (p. Aktau, Karaganda region) Modifying additives Movilit DM 207 2P and tylose MB 15009 P2 from Clariant (Germany).

To identify characteristics of compounds included in the original tailings, cements and functional additives, following were used: the DRON-ZM X-ray unit, the MIN-8 polarizing microscope, the MOM-1500 derivatography unit, and the Specord M-80 infrared spectrophotometer.

Preparation of samples for these devices was carried out according to instructions for their manufacture for analysis in the appropriate installation (Shestoperov 1991; Dauletov et al. 1985; Suleimanov et al. 1987).

X-ray diffraction patterns were obtained on a DRON-ZM diffractometer. Polycrystalline samples were carefully ground to a powder and placed in a plexiglass cuvette, the bottom of which was smeared with a thin layer of Vaseline. To avoid texturing, the sample was cut off with a piece of glass. The X-ray tube with a copper anode operated in the mode: 40 kV and 20 mA, the speed of the detection unit was 1 °/min. Shooting interval 2θ from 4 to 64 °.

Samples for examination in the Superprob-733 scanning electron microscope were prepared as follows. Fresh chip of the sample was made. Then samples were placed in a vacuum installation, where using ion sputter, fractures of sample were covered with a thin structureless layer of gold (to improve the yield of secondary electrons when studying samples in scanning electron microscope). Using this microscope, secondary electron images (SEI) were obtained from the surface of the chips at various magnifications.

Testing of DBM components was carried out in accordance with the relevant regulatory documentation for these materials.

Properties of cement that meet GOST 10178-85 "Portland cement and Portland slag cement. Technical conditions" were determined according to GOST 310. 1-76 - 310. 4-81 "Cements. Test methods".

Since concentration tailings under study are, as shown above, raw materials that do not require additional processing, they can be considered as the main component that meets the requirements according to GOST 8736-93 "Sand for construction work. Test methods".The classification of used additives of organic and inorganic origin was carried out according to GOST 24211-91 "Additives for concrete. Technical conditions".

Traditionally, DBM is produced by joint grinding of binders (cement, lime, gypsum) and mineral components (sand, carbonates) with modifying additives. However, such mixtures do not provide required construction and technical characteristics that meet requirements for materials for repair and restoration work. To increase homogeneity of resulting mixtures and improve workability, we have developed method for producing mortar mixtures from dry components.

Mortar mixture for testing should be prepared as follows: water and cement are mixed until indiscrete mass, and then the calculated amount of tailings and modifying additives are added. Resulting mortar mixtures are placed in detachable steel molds in the form of a cube (3 control samples measuring 7 × 7 × 7 are prepared for strength testing) and 3 samples in form of rectangle measuring 4 × 4 × 16 (for bending testing). After this, mortar mixture in molds is subjected to vibration on a laboratory vibration platform for 1 min (GOST 5802-86). After a day, the samples are removed from the molds and placed in a storage chamber in which the specified temperature and humidity are maintained. Testing of samples is carried out at certain time intervals (1…28 days). After 28 days, concrete grade (class) is determined based on strength indicators.

Study of influence of individual additives with different concentrations, as well as their complex content on processes of hydration and binder hardening was carried out on samples 2 × 2 × 2 cm. After storing samples under normal conditions at certain time intervals of 1, 3, 7, 14 and 28 days, cubes strength was determined according to GOST 10180-90: "Concrete. Methods for determining strength using control samples on a press type IP-100.

Hydration feature of the binder with various additives and qualitative composition of hydrated cementing phases over time were monitored by methods of derivatography, X-ray diffraction and other analyses. Setting time of cement in combination with various additives and their quantities was determined using the standard method for determining the properties of cement according to GOST 310.1-76 "Cements. Test methods".

To study the physical and mechanical properties of mixtures using different fillers (aggregates) with the addition of various additives, samples with dimensions of 7 × 7 × 7 cm were molded. In parallel, such properties of mixtures as the average density of dry mixture and mixture with certain mobility, water-holding capacity, humidity, water absorption were determined, compressive strength, frost resistance according to GOST 5802-86 "Building mortars. Test methods" and according to KRST 1168-2006 "Dry mixtures for construction. Technical conditions".

Mobility determination of solution mixture was carried out on standard cone when it was immersed in mixture. Moisture content of mixture was determined after drying sample in oven to constant weight.

Water absorption of samples was determined after they were saturated with water to constant weight in container and then dried in oven.

Water-holding capacity of solutions was determined by testing layer of the mixture laid on blotting paper that absorbs moisture from the solution. The strength of mortar samples with dimensions of 7 × 7 × 7 cm after 1, 3, 7, 14 and 28 days of hardening and under normal conditions was determined using an MS-500 press in accordance with GOST 10180-90 "Concrete. Methods for determining strength using control samples."

Frost resistance was determined by the standard method by alternately freezing and thawing samples measuring 7 × 7 × 7 cm.

After processing the test results, curves were constructed for the dependence of strength on the hardening time of the samples, the amount of added additives and their complex content. Processes of structure formation were studied by determining the plastic strength in cement + additive + water systems using a conical plastometer with a cone angle of 30°. Working compositions were prepared in quantities of about 100 g, placed in a container of a certain capacity and at intervals of 20–30 min. Measured the amount of immersion of the plastometer cone into the hardening mixture. Having calculated corresponding values of plastic strength, graphs of its dependence on time were constructed.

3 Conducting Laboratory Studies of Obtaining New Compositions of Building Materials Based on Current Copper Ore Enrichment Tailings

For the preparation of dry building mixtures (DBM), following raw materials were used:

- tailings from the Zhezkazgan enrichment plant;
- Portland cement PC 400 D20 of JSC "Central AZIA Cement" (Aktau village, Karaganda region) and Portland cement PC 400 D20 of the Shymkent cement plant;

- functional additives: movilit DM 2072R and tylose MV 15009 P2 from Clariant (Germany);
- granulated blast furnace slag of the Karaganda metallurgical plant—as a hydraulic additive in the production of Portland cement PC 400 D20;
- semi-aqueous building gypsum grade G10.
- enrichment tailings Zhezkazgan enrichment plant. Chemical composition %: SiO_2—76,02; CaO—3,56; Al_2O_3—10,45; MgO—1,17; Na_2O—2,62; K_2O—1,93; TiO_2- 0,33; MnO-0,04; CuO-0,16.

Particle size distribution of tailings, %: particles larger than 0, 63 mm make up 0.05%; particles smaller than 0,63 mm, but larger than 0,315 mm—5,09%; particles smaller than 0,315 mm, but larger than 0,16 mm—32.96%; particles smaller than 0,16 mm—61,9%.

Fineness modulus—0,43. Consequently, the tailings from the enrichment of ZhezOF are very fine-grained sand with a high content of SiO_2. The physical and mechanical properties of Portland cement PC 400 D20 of Central AZIA Cement JSC are presented in Table 1.

Data in the table shows:

- grinding fineness as it passes through sieve № 008 reaches 95,5%.
- normal density is 27,2%;
- cement cake in boiled water maintains uniform volume changes;

Table 1 Physical and mechanical properties of cement ПЦ 400 Д20 (JSC «Central AZIA Cement»)

Indicator name	Regulatory document for test method, product number	Norm according to Regulatory document 10178–85	Actual meaning
Grinding fineness as passed through sieve № 008, %	GOST 310.2-76	no less 85	95,6
Normal thickness, %	GOST 310.2-76	–	27,2
Setting time, hour/min:			
- start	GOST 310.3-76	not earlier 0,45	1–10
- end		not later 10–00	3–20
Strength after steaming, kgc/cm^2 (MPa) at:			
- bend	GOST 310.4-81	-	41,3(4,13)
- compression		-	279,8 (27,98)
- meaning			
Strength at the age of 28 days, kgc/cm^2 (MPa) at:			
- bend	GOST 310. 4-81	no less 55	66,5 (6,65)
- compression		no less 400	415,6(41,56)
Uniformity of volume change (in boiled water)	GOST 310.3-76	there should be no cracks	withstood

Table 2 Physical and mechanical properties of cement PC 400 D20 (JSC Shymkent Cement Plant)

Indicator name	Regulatory document for test method, product number	Norm according to Regulatory document 10178–85	Actual meaning
Grinding fineness as passed through sieve № 008,%	GOST 310.2-76	no less 85	90,0
Normal thickness,%	GOST 310.2-76	–	28,7
Setting time, hour/min:			
- start	GOST 310.3-76	not earlier 0,45	1. 50
- end		not later 10–00	2. 58
Uniformity of volume change (in boiled water)	GOST 310.3-76	there should be no cracks	withstood
Strength after steaming, kgc/cm^2 (MPa) at:			
- bend	GOST 310.4-81	–	31,5 (3,15)
- compression		–	233 (23,3)
Strength at the age of 28 days, kgc/cm^2 (MPa) at:			
- bend	GOST 310. 4-81	no less 55	55,6 (5,56)
- compression		no less 400	401,3 (40,13)

- after steaming the samples, their bending strength reaches 41,3 kgc/cm^2, and compressive strength – 279,8 kgc/cm^2;
- after 28 days of hardening under normal conditions, compressive strength is 415,6 kgf/cm2.

These data indicate that this Portland cement PC 400D20 meets requirements of regulatory documents.

Physical and mechanical properties of Portland cement PC 400 D20 ShTsZ are given in Table 2.

From the given data it follows that:

- grinding fineness as it passes through sieve № 008 reaches 90,0%;
- normal density is 28,7%;
- setting time, hour/min: beginning—1-50; end—2-58;
- uniformity of volume change when boiling in water—withstands without formation of cracks on samples;
- samples strength after steaming in bending is 3,15 MPa, in compression—23,3 MPa;
- samples strength at the age of 28 days reaches 5,56 MPa, and in compression—40,13 MPa.

4 Method for Producing Energy-Efficient Dry Building Mixtures Based on Modified Diatomite

4.1 Current Trends in the Development of Dry Building Mixtures Technology

In modern construction practice, more and more attention is paid to the introduction of new effective materials for the construction and repair of buildings and structures, which can significantly reduce labor costs at the construction site and reduce the consumption of materials while simultaneously improving the quality of the work performed. Such materials include dry building mixtures (DBM), which today are a product of high-tech technologies. Depending on their purpose and property requirements during their manufacture at factories, the composition and characteristics of building mixtures are easily changed by introducing a complex of various additives. Dry building mixtures are considered more effective materials compared to traditional cement-sand mortars of centralized preparation.

Dry building mixtures, ready for use, are widely used in finishing works. In the Kazakhstan market, dry building plaster mixes based on lime are represented by the brands “Alinex”, “Alina”, “Knauf” and other suppliers. All of them have a fairly high cost. In the structure of the price of dry building mixes, the cost of raw materials is from 30–40%. In this regard, use of local materials in the formulation of dry building mixtures s is relevant.

Scientific project (Project 2023) is devoted to the development of a technology for obtaining energy-efficient dry building mixes using local raw materials that are underused in dry mix technology. Kazakhstan has significant reserves of diatomite that can be used in the production of energy-efficient dry building mixes. Thus, in a number of CIS countries, diatomite deposits are actively developed for the use of this raw material in the national economy, including in technologies for obtaining building materials and products. Diatomite, consisting mainly of amorphous silica, in a finely dispersed state interacts with binders in the presence of moisture, but the strength of such solutions during air-dry hardening is low. It is necessary to develop technologies and scientific and technical solutions aimed at modifying diatomite and solving a number of problems associated with a significant reduction in cost and the development of an alternative method for increasing the activity of diatomite and obtaining dry building mixes based on it. Solution to the above problems will reduce energy costs during diatomite activation and create dry building mixtures, coatings based on which will have increased operational resistance.

Dry building mixtures occupy special place among building materials, since they have several advantages and are highly effective in technical and economic terms, so they give special architectural expressiveness and aesthetics to urban planning, not to mention creation of favorable living conditions and rational use of energy resources. Trend in technology development for production of dry construction mixtures shows influence of two main factors that accelerate and inhibit development of this production. Thus, volume of production of dry building mixtures in

Kazakhstan will continue to increase in the near future, associated with construction of civil and residential facilities, which predetermines increase in demand for dry mixtures, and also relevant factor in increasing market for dry mixtures is its replacement with competitive local products.

Dry building mixtures are fine-grained compositions that contain mineral binders, fillers of certain fractions and chemical additives (Viktorovich 2000; Fedulov 1999; Bezborodov 1998). The most common components and materials used in the production of dry mixes are presented in Table 3.

Analysis of literary sources indicates that in Western European countries up to 65% of total volume of dry mixes produced is accounted for by finishing plaster compositions. In addition to trowelling compounds, one of the promising areas is production of putty dry mixtures (Bolschakow 1997; Über die Welt 2008).

Leading companies “Knauf”, “PFT”, “M-Tek” in production of dry putty coatings have developed cement-sand and cement-lime-sand dry mixtures containing Portland cement, fluff lime, construction quartz sand and multifunctional chemical additives. As multifunctional chemical additives, multicomponent compositions are used, including cellulose ethers, surfactants, stabilizers and inhibitors.

According to the authors (Vasilik 2002; KNAUF 2003) the Waker KG company specializes in production of dry mixtures for making facade plaster coatings. Company produces 11 types of dry mixtures for various purposes for making plaster coatings on brick surfaces, creating coarsely bumpy surfaces applied by spraying, as well as thin coating coatings. Solutions prepared from this mixture have accelerated hardening, since mixture contains additives that intensify processes of setting and hardening of composition.

Currently, production and consumption of dry mixtures in Kazakhstan continues to be in stage of dynamic development. Large projects with foreign participation for production of dry mixtures have been implemented (Meshkov 2000).

The data presented indicate that competitive field is being formed in the Kazakhstan market from leading Western and domestic producers of dry mixes, including those producing dry mixes under license. However, for many domestic producers of dry mixtures engaged in their own developments, fundamental issue of ensuring competitiveness remains increasing their quality and uniformity of properties. Domestic manufacturers solve such problems by introducing chemical additives (modification), as one of the most effective and simple ways to change the properties of building materials and improve their quality. Moreover, foreign

Table 3 Components and materials for production of dry mixtures

Binders	Fillers	Additives
Portland cement, white cement, gypsum, lime, anhydrite, aluminous cement, dispersible polymer powders	Quartz sand, chalk, limestone, dolomite, microsilica, kaolin, perlite, fly ash, fiber (fiber), light aggregates (pumice, expanded clay, etc.), pigments	Plasticizers, stabilizing and water-retaining dispersible polymer powders, retarders, accelerators, thickeners, pore-forming and antifoaming additives, hydrophobisators

companies are actively offering various chemical additives, allowing domestic companies to produce modified dry mixtures in a fairly wide range.

Good trend in modifying dry building mixtures is use of reinforcing fibers of both organic and inorganic origin. Reinforcement of dry building mixtures significantly increases tensile and bending strength, compensates for disadvantages of fractional composition, reduces shrinkage, and improves thixotropic properties.

It is no secret that, for all their advantages, modified dry mixtures are much more expensive than a mixture of mineral binder and sand. Chemical additives in simple formulations, for example, in tile adhesives for interior work, cost about 80% of all raw material costs, and in complex ones—up to 97%. Therefore, desire to save on mineral components is wrong. On the contrary, mineral composition must be selected in such a way that consumption of chemical additives is minimal.

The most widely used in production of dry mixtures are plasticizing additives, mainly those that are produced in form of solid powder (Goreglyad 2001). For production of dry mixtures, additives based on lignosulfates (LST, PLS-1P, PLS-N, LSTM-2, LTM-P), which are waste from pulp and paper industry are widely used. Plasticizer LST (technical lignosulfate) is currently produced by Almaty, Karaganda and Kostanay pulp and paper mills of the Republic of Kazakhstan.

Plasticizer PLS-1P is produced on the basis of modified LST lignosulfate; it does not contain free acids and volatile sulfur compounds. Additive in powder form mixes well with water, due to modification and purification it does not have toxic properties and is identical in its chemical actions to PLS-I.

The multifunctional additive LTM-P is a powdery material based on LST, modified with sodium and calcium salts of mineral acids and surfactants. Sulfate-yeast mash (SYM) is used as plasticizers in combination with slaked lime, sodium chloride, ferrous sulfate, polyphosphate or LST lignosulfate. These additives are effective retarders for gypsum (Demyanova 2001). For the same purpose, ethylenediamine, nitrilotrimethylphosphate polyalkylamide, acid and potassium tartrate are added to compositions of dry mixtures. As a complex additive in composition of plaster mixtures, sodium humate reagent can be used—highly effective powder additive, which is product of chemical–mechanical processing of brown coal.

Analysis and generalization of advanced foreign and domestic experience in production of dry mixes revealed important feature regarding composition of dry mixes. Recipe for most dry mixes includes: cement (universal binder for exterior and interior work), gypsum (binder used only for interior work), fine aggregate (in most cases, sand dispersed into fractions), inorganic chemical and polymer additives. It is noteworthy that foreign and domestic manufacturers most often use cement or gypsum as a binder. Lime as a binder for dry mixes is used relatively rarely.

Lime is mainly introduced into formulation of dry cement-based mixtures as improving mineral additive. From economic point of view, use of high-quality cements (M400, M500 and more) is not always justified, especially in cases where composition of dry mixture is not designed to withstand large compressive loads. Example is plaster compositions, where material applied to surface experiences tensile forces and service life of traditional finishing compositions based on cement (or gypsum) is largely limited by their low crack resistance. In plastering and decorative compositions, it is most advisable to use low-quality formulations that are less durable, but capable of withstanding tensile forces.

Competitiveness and profitability of production depend on cost of production. Today, development of high-quality and inexpensive dry mixtures has become relevant for domestic producers of dry mixtures. Reducing cost of dry mixtures can be achieved not only by reducing the content of chemical additives. Promising direction in the production of dry mixtures is use of binders and fillers based on local rocks. Local raw materials, due to reduction in transportation costs are much cheaper than those imported from other regions. This opens up the possibility of integrated development of local raw materials, while significant economic effect is achieved from use of dry mixtures, which will ultimately help reduce construction costs.

Studies discussed above indicate possibility of creating a low-quality binder based on a mixture of lime. Thus, it becomes possible to create recipe for economical dry mixture intended for plastering and finishing work.

Currently, there are various types of chemical additives on the domestic market, supplied from abroad and allowing production of modified dry mixtures for finishing work in fairly wide range. First of all, these are cellulose esters, which increase the water-holding capacity of gypsum and cement mortars, improve their workability and workability, and have positive effect on the rheological properties of mortars. Due to additional water retention, cellulose ethers provide thickening effect, thereby creating conditions for more uniform and complete hydration of mineral binders in thin-layer finishing coatings.

Currently, cellulose esters with a high degree of esterification are widely used as part of modified dry construction mixtures. Most often, methylhydroxyethylcellulose of various brands is used for these purposes. Considering degree of modification of cellulose ethers, it is possible to predict such properties of solutions based on dry mixtures as water-holding capacity, viscosity, water requirement, and fixing ability when used in thin-layer finishing technologies. Along with the above-mentioned cellulose ethers, other types of water-retaining additives have appeared on the market, for example, methylhydroxypropylcellulose with a low degree of esterification. These types of cellulose ethers can also be used in dry mixtures for finishing work in order to reduce their cost, however, when selecting compositions, their greater hydrophilicity and lower water-holding capacity should be taken into account, which in some cases can affect the technological properties of solutions.

A higher level of modification is achieved by introducing redispersible copolymer powders (RCP) into composition of dry construction mixtures, which can act as additive or independent polymer binder. Mineral binders provide high compressive strength and have poor adhesion to unrelated materials, especially those that do not absorb water—glazed ceramics, plastics, metals, polystyrene foam, etc. When dosing redispersible copolymer powders from 0.5 to 5 wt.%, the processability of mixture, adhesion to the base, bending strength, water resistance, and frost resistance are significantly improved. At a dosage of more than 5 wt.%, a significant increase in the plasticity of the mixture occurs, and RSPs begin to work as independent polymer binders. Therefore, mixtures modified by them begin to demonstrate elastic properties: they can withstand significant deforming loads and have increased abrasion resistance.

Wide variety of polymer binders and special additives designed to modify properties of dry construction mixtures are offered by the Rodna company of the largest chemical and pharmaceutical concern Ron-Roulenc.

Unfortunately, market for Kazakhstani cellulose ethers, redispersible latex powders, and superplasticizers is limited. Na-CMC and cellulose ethers are produced by only three enterprises. These enterprises produce limited range of cellulose derivatives CMC, methylcellulose and Na-CMC, and of very low quality. Thus, Na-CMC is produced with a mass fraction of main substance up to 50–64%. Dynamic viscosity usually does not exceed 150–200 MPa, which does not allow the use of Na-CMC in the main range of dry construction mixtures.

Cost of domestic products is significantly lower than foreign ones, but they can be used in isolated cases after appropriate technical and economic justification. In this regard, domestic manufacturers of dry mixtures are focused on use of foreign-made polymer additives.

Currently, in the construction market of Kazakhstan increase in production volumes of domestically produced dry mixes is observed, quality of which is not inferior to foreign analogues. However, most of the products are produced using imported components, which inevitably leads to higher prices for dry mixtures. While water-soluble polymer additives are almost always imported, there are high-quality aggregates and fillers in Kazakhstan.

Material costs account for about 40% of the total cost, and their reduction in price is one of the main ways to reduce costs in modern construction. Therefore, promising direction in production of dry mixtures is use of mixed binders modified with finely ground dispersed fillers based on local soft and loose rocks. These binders are much cheaper than cement binders, which makes it possible to organize comprehensive development of local raw materials, while achieving significant economic effect in use of dry mixtures in.

Currently, considerable experience has been accumulated in the study of fairly pure varieties of siliceous rocks (diatomites, tripoli, opok), their classification has been developed, their chemical composition, physical and technical properties are known (Yestemesova et al., 2019; Kozlov 2000).

According to geological prospecting data, siliceous rocks within the Aktobe region of the Republic of Kazakhstan are widespread. Most of the surface of the areas is occupied by siliceous Paleogene deposits, forming numerous deposits of diatomites. The depth of diatomite is 0.5–30 m. The thickness of diatomite in outcrops is 7–20 m.

Diatomite is loose or weakly cemented siliceous sedimentary rock, consisting primarily of shells of diatoms, and has white, gray or pinkish color. In their natural state, diatomites have large, evenly distributed porosity, reaching 80–85%. Nature of the pores is closed and small. Chemical composition in its pure form can be expressed by the formula $mSiO_2nH_2O$. Silica in diatomites is in amorphous state and amounts to 78–95%. In addition to silica, diatomites contain small amounts of calcium and magnesium carbonates and some other impurities.

Diatomite is used to produce lime-diatomite binders, effective low-burning materials, lightweight concrete aggregates, ceramic products in the form of shells and segments, etc.

Sufficient reserves of siliceous rocks in the Aktobe region of the Republic of Kazakhstan make it possible to organize production of effective dry construction mixtures.

4.2 *Influence of Dry Mixture Filler on Structure Formation of Lime-Siliceous Compositions*

Properties of dispersed materials, widely used in formulation of dry mixtures are determined mainly by state of their surface Any real surface of a solid body, being in itself one of the main defects in the three-dimensional structure of crystal lattice, is inhomogeneous in geometric, chemical and electronic sense. Breakage of the periodic lattice leads to change in coordination sphere of surface atoms and rehybridization of bonds (Altayeva et al. 2019, 2020).

Disturbance in the course of periodic potential on surface, caused by the break of periodic lattice, presence of various macro and microdefects, as well as various surface chemical compounds, leads to appearance of a wide range of surface states—adsorption centers. Nature of energy heterogeneity of surface determines its chemical properties, specifying spectrum of active surface centers, which can be both surface lattice atoms and surface functionals formed as a result of interaction with environmental molecules.

Initial stage of interaction between solid powder and liquid is adsorption. It is important to note that not the entire surface, but only active centers take part in adsorption. Adsorption by donor-acceptor mechanism not only leads to surface modification, but also affects the charge state of surface. Most of the processes occurring with participation of surface of solids are local in nature and are largely determined by energy parameters of specific active centers. Surface of solid has properties of acids and bases. By solid acid (base), a solid body is supposed on which change in color of main (acid) indicator occurs, or chemical adsorption of base (acid).

Based on fact that surface of silica contains mobile proton capable of ion exchange, it cannot be inert substance, but is active component of system and possible catalyst for hydrate formation. In works (Patent No. 2776, 2018; Kolesnikova et al. 2019) surface of silica was studied in more detail, which is more often than other materials used aggregates and fillers for composites. Significant step in resolving issue of nature of active centers on surface of silica was establishment of important role of structural hydroxyl groups in adsorption processes. It was shown (Patent No. 2776) that surface of silicate particles under normal conditions is covered with OH groups, and these groups are centers on which water molecules and many organic molecules are adsorbed. Silicon atoms on surface of amorphous silica are not exactly in regular geometric order, so hydroxyl groups attached to silicon atoms are located at different distances from each other. Consequently, they are not equivalent in their behavior with respect to adsorption phenomena or chemical reactions.

Several works (Loganina 1999; Loganina and Puchkov 2002) have shown that silica surface exhibits proton-donor activity. Proton centers play significant role in variety of physical and chemical phenomena. Strength of hydrogen bonds depends on proton-donor properties; proton (Brønsted) centers are directly involved in acid–base reactions and oxygen exchange.

Data obtained to date on nature and patterns of reactions involving silica surface make it possible to modify surface to improve its adsorption and ion-exchange properties, and to make surface hydrophobic or hydrophilic.

For example, in (Ermilov 1987) it was found that with partial dehydration of silica surface, its reactivity increases due to increase in number of isolated OH groups.

Reactivity of silica is function of surface state, modification of which leads to change in nature and concentration of active centers. It is known that diatomite, consisting predominantly of amorphous silica in finely dispersed state in presence of moisture, interacts with lime-cement binders, but strength of such solutions during air-dry hardening for month is low.

Therefore, it is necessary to strengthen interaction of diatomite with lime-cement binders, for which surface activity of filler should be increased. However, effect of changes in chemistry of diatomite surface and increase in its reactivity during modification process has not been studied.

Analysis of literature data has shown that issues of modifying surface of fine mineral fillers of dry mixtures and development of finishing compositions based on them remain very pressing task today.

In this regard, goal of work is to develop recipe and technology for preparation of dry lime-diatomite mixtures for finishing building products and structures.

4.3 Study of Physical and Mechanical Properties of Dry Construction Mixtures Based on Modified Diatomite

There are several known methods for activating diatomite filler, one of which is thermal method, which was adopted in work, as it has a number of positive advantages.

Modification of diatomite filler was carried out using thermal method and temperature range was taken from 20 °C to 950 °C. Based on results of heat treatment, tests were carried out on resulting raw materials with fixation of compressive strength indicators at the age of 28 days on samples 20 × 20 × 20 mm and at a ratio of 1:3, Table 4.

Table 4 Compressive strength depending on the heat treatment temperature of diatomite

Processing temperature, °C	Aged compressive strength 28 days, MPa
20	0,88
200	0,91
300	0,92
650	3,20
950	3,56

Table 5 Compressive strength of lime-diatomite compositions

Lime-diatomite ratio	Additive type	Additive content, %	Aged compressive strength 28 days, MPa
1:1	–	–	0,3
1:2	–	–	0,4
1:3	–	–	0,5
1:3	PFM-NLK	0,5	0,8
		0,75	0,8
		1,0	0,8

Analysis of the data given in Table 4 indicates that heat treatment of diatomite at low temperatures (200 °C and 300 °C) does not have significant effect on compressive strength values of compositions. Increasing firing temperature to 650 °C leads to increase in strength characteristics to Rcom = 3.20 MPa. However, the greatest effect is achieved by heat treatment of diatomite at a temperature of t = 950 °C. Value of ultimate compressive strength is Rcom = 3,56 MPa. At firing temperatures of 650 °C and 950 °C, diatomaceous earth acquires a bright orange tint, which make it possible to diversify color range of finishing layer without introducing pigments. However, from point of view of energy consumption, heat treatment of dry mixture filler at t = 650 °C is more expedient.

Significant increase in strength characteristics of compositions based on thermally treated diatomite indicates that decisive influence on interaction with lime is not the chemical and mineralogical composition of diatomite, but number of active surface centers.

Intensity of hardening and final strength of mortars largely depend on its composition, hardening conditions, as well as type and amount of additives used.

Data were obtained on compressive strength of lime-diatomite compositions at different ratios of dry mixture components.

Analysis of data in Table 5 shows that samples with higher content of diatomite (fine fraction) have higher compressive strength. Thus, compressive strength of samples with a lime: diatomite ratio of 1:1 was 0,3 MPa; with a ratio of 1:2–0,4 MPa, and with a lime-diatomite ratio of 1:3—from 0,5–0,8 MPa.

It was revealed that an increase in the dispersion of the dry mixture filler leads to an increase in the strength characteristics of lime-diatomite compositions. The compressive strength of samples containing finer ground diatomite was 0.8 MPa.

During determining compressive strength of lime-diatomite compositions, it was noticed that values obtained when measuring this indicator are not the same. Strength values are influenced by heterogeneity of starting materials, accuracy of dosing of components, preparation thoroughness of dry mixture and other factors. In this regard, to assess reliability of experimental results obtained statistical data processing was carried out.

To assess homogeneity of finishing compositions, results of control tests of lime-diatomite samples during hardening process were studied. Calculated data indicate that highest values of the coefficient of variation (V = 8–12%) correspond to samples tested at the age of 7 and 14 days. Thus, at the age of 7 days, coefficient of variation for composition lime: diatomite 1:3 was V = 12,06%, while composition containing thermally treated diatomite as a filler had coefficient of variation V = 7,86%. Mathematical processing of experimental data at the age of 28 and 90 days shows a decrease in values of coefficient of variation, which indicates increase in quality of homogeneity.

5 Technology Development of Water Dispersion Paints on Acrylic Base with Content as Filler Modified Diatomite

5.1 Current State of Development of Paints and Varnishes

Production of high-quality paints and varnishes that meet all modern requirements is impossible without constant improvement of formulations and search for new solutions to reduce production costs and at the same time preserve valuable properties of the material. Due to constant rise in cost of raw materials, as well as strengthening environmental requirements, use of cheap but effective components in paints and varnishes that can replace expensive or toxic raw materials is becoming increasingly important. To reduce cost of paint and varnish products, it is advisable to use local raw materials as fillers.

In production of paints and varnishes, mineral fillers are used as inexpensive components to reduce production costs. By selecting fillers, it is possible to significantly improve such characteristics of paints and varnishes as adhesion, moisture resistance, viscosity, pouring, weather resistance, mechanical strength, hardness, etc.

American scientists have presented anti-corrosion paint and varnish composition to provide anti-corrosion protection for metal substrates. Composition includes from 20 to 97% (by weight) of filler. Filler includes at least one material with a hollow structure, in which cavity is filled with organic or inorganic inhibitor or antioxidant. Such a filler can be diatomite, zeolite, or carbon. Long-term anti-corrosion protection is achieved by dosed release of inhibitor from filler cavity during operation of paintwork.

The most well-known and long-used filler for paint systems is chalk. Kazakhstan has large reserves of diatomite, which makes it possible to use them in paint and varnish coating compositions.

According to geological prospecting data, siliceous rocks within the Aktobe region of the Republic of Kazakhstan are widespread. Most of the surface of the areas is occupied by siliceous Paleogene deposits, forming numerous deposits of diatomites.

Diatomite is a loose or weakly cemented siliceous sedimentary rock, consisting primarily of shells of diatoms, and has a white, gray or pinkish color. In their natural state, diatomites have large, evenly distributed porosity, reaching 80–85%. Nature of pores is closed and small. Chemical composition in its pure form can be expressed by the formula $mSiO_2*nH_2O$. Silica in diatomites is in amorphous state and amounts to 78–95%. In addition to silica, diatomites contain small amounts of calcium and magnesium carbonates and some other impurities.

During solving several environmental problems in construction industry, water-dispersed coatings, production, and use of which are not associated with use of toxic and fire-hazardous organic substances, are becoming increasingly important. Their main advantages include low toxicity, quick drying, ability to paint wet surfaces and carry out painting work at high air humidity.

During working with water-dispersion paints and varnishes, requirements for labor protection, fire and explosion hazards of painting work are reduced, and use of explosion-proof equipment for production and application is not required.

Thus, use of water-dispersed paints and varnishes allows not only to save on cost of irretrievably lost solvents, ventilation, and safety measures, but also on ensuring the fire safety of process.

5.2 *Technologies Research for Producing Polymer-Based Composites Using Various Fillers*

Increasing efficiency of building materials, that is, improving existing ones or giving them new properties, is one of urgent tasks of modern developing production.

In recent years, demand for materials intended for their finishing, including paint and varnish coatings, has increased; it is advisable to use local raw materials as fillers.

Analysis of DIY segment of market for finishing building materials for interior work, carried out by the authors in the Republic of Belarus and the Republic of Lithuania, showed that water-dispersion paints for universal use are dominant (Khaletskaya 2014). Among surfaces recommended for painting, manufacturers often indicate entire possible set, for example: "mineral (brickwork, concrete, aerated concrete, plastered or putty surfaces, cement-fiber boards, plasterboard, etc.), wood, previously painted with water-dispersion paints and other porous surfaces of buildings and structures." Additionally, potential applicability of product for painting textured substrates, such as structural wallpaper on paper, vinyl, and non-woven backing, is indicated.

At the same time, versatility of considered water-dispersed materials in some cases can lead to significant problems. For example, during painting structural wallpaper, it is necessary to preserve its relief, which forms visual perception of this material. Conventional paints with coarse and medium fillers can "clog" the details of the wallpaper surface, forming a rough, sloppy coating. In this regard, thin-layer water-dispersion paint for interior work was developed. Authors developed and tested in industrial conditions formulation of thin-layer paints for interior work based on styrene acrylic dispersion. Approximate recipe is given in Table 6.

Test paint batches were made using a laboratory dissolver with a fixed cutter rotation speed of 900 rpm in polymer container in one stage. When determining content of components, transferring convenience of recipe for industrial production was considered.

Calculated volume pigment concentration (VPC) for composition is 33.21%, calculated mass fraction (MF) is 58.49%. VPC value for composition is significantly less than critical VPC, which corresponds to glossy and semi-gloss paints with a high content of film-forming agent.

Recipe uses rutile titanium dioxide, obtained by sulfate method as a white pigment. Surface of pigment particles is treated with aluminum and zirconium oxides, because of which it is resistant to photodestruction and is also distinguished by high

Table 6 Approximate formulation of thin-layer water-dispersion paint

№	Component name	Content, mass. %
1	Water	15,74
2	Propylene glycol	2,95
3	Calcite (marble powder), fraction 0.9 microns	18,44
4	White pigment (titanium dioxide)	15,37
5	Polymer film former (aqueous copolymer styrene acrylate dispersion)	44,26
6	Hydroxymethylcellulose (rheological additive)	0,10
7	Non-ionic associative thickener (rheological additive)	0,37
8	Coalescent (Dalpad Filmer™, Dow Chemical Europe)	1,57
9	40% aqueous sodium polyacrylate solution (dispersant)	0,39
10	Non-ionic wetting agent based on tridecyl alcohol	0,15
11	Mineral oil based defoamer	0,25
12	Container preservative	0,39
13	Ammonia (acidity regulator)	0,02

whiteness with a very slight yellow tint. Filler in system was calcite filler Omyacarb CalciglossTM (Omya, Gummern) with median particle size of 0.9 microns.

Injection of non-ionic wetting agent into formulation is dictated by the need for subsequent tinting of the paint when it is used as a white base in computer tinting systems. Combination of thickeners allowing to achieve optimal painting characteristics of material. Paint composition after testing properties and formulation approval was reproduced in industrial environment. Weight of one batch was approximately 3000 kg.

Research of resulting thin-layer paint was carried out, and both paint itself and coating based on it were examined. Test methods complied with technical regulations in force in the paint and varnish industry. Research results are presented in Table 7.

High content of film-forming agents in paint formulation makes it possible to achieve high coating resistance to static effects of not only water, but also, for example, disinfectant solutions based on quaternary ammonium compounds, which makes it suitable for painting walls in healthcare institutions. At the same time, coating resistance to wet abrasion increases, and its class according to STB EN 13300-2011 decreases. At the same time, a large amount of polymer binder requires careful selection of coalescent to prevent residual stickiness of coating.

Modern paints and varnishes are highly filled systems. According to the European standard, regulatory standard for indicator "dry matter residue" is a value of at least 75% for paints and at least 97% for plastics (Shcherbakova et al. 2013).

Table 7 Characteristics of developed thin-layer water-dispersion paint

№	Indicator name	Test method	Actual value
1	Appearance of coating	GOST 28196	Smooth and uniform silky surface
2	Mass fraction of non-volatile substances, %	GOST 17537	58,3
3	pH value	GOST 28196	7,6
4	Covering power of dried film, g/m^2	GOST 8784	1
5	Grinding degree, μm	GOST 6589	25
6	Coating resistance to static action of water at a temperature of (20 ± 2)°C, h	GOST 9.403	72
7	Drying time to degree 3 at temperature (20 ± 2)°C, h	GOST 19007	1
8	Paint viscosity Brookfield, 20 rpm, cP		4220
9	Paint viscosity according to ICI, 750 rpm, cP		47
10	Optical hiding power, %		95,61
11	Color in Lab coordinates		L 96,06 a −0,58 b +0,68

For development of marking paint, the most important indicator of quality of materials is their service life (Brock 2007).In this regard, following requirements are imposed on coatings being developed: components for preparing coating must be non-toxic; deviations in the dosage of components within the limits of up to 5% from calculated values should not significantly affect physical and mechanical properties of coating; in appearance, coating before application should be a homogeneous mobile liquid without sediment or delamination, and after application it should form a smooth, single-color surface; coating density should be in the range of 1,2–1,6 kg/m^3; the content of non-volatile substances must be at least 70%; the coating viscosity should be 60–120 s-1; the degree of grinding should be no more than 70 microns; Drying time of coating at a temperature of 20 ° C and a relative air humidity of no more than 75% should be (Yakovlev 2008):

1) to degree 3 – no more than 30 minutes;
2) to degree 5 – no more than 50 minutes.

Coatings based on pure acrylic paints and varnishes with a high content of anti-corrosion pigments are characterized by unique durability and chemical resistance. Coating, consisting of a zinc-enriched acrylic primer and acrylic varnish, has a service life in particularly aggressive environments of over 15 years (Kazakova 2003).

As a result of numerous experiments, interior paint that was resistant to washing was developed, and a large percentage of consumption of polyvinyl acetate dispersion per unit of product made it possible to produce paint without addition of coalescent, with unchanged physical and mechanical properties.

Considering the previously described features of composition, during tests following optimal formulation of water-dispersion interior paint based on polyvinyl acetate dispersion was identified; components of this development are summarized in Table 8 (Sokov et al. 2019).

Deteriorating environmental situation in the world is causing stricter requirements not only for various industries, but also for the materials they produce, since even newly developed products are not always environmentally friendly for humans and the environment. Among such materials, protective polymer coatings play an important role. It is known that currently the most common and accessible way to

Table 8 Recipe for water-dispersion paint

Components	Contents in %
Water	16,00–22,00
Preservant	0,10–0,30
Defoamer	0,10–0,50
Filler	20,00–30,00
PVA dispersion	30,00–35,00
White pigment	13,00–18,00
Silicone additive	0,50–1,50
Modifying additives	0,50–1,00

solve this problem is the use of filled water-based acrylic compositions (Skorokhodova 2013; Pavlov 2012).

Most manufacturers use chalk, talc and microcalcite as fillers, which do not provide sufficient protective properties. As is known, modifying targeted additives and natural fillers play an important role in the production of protective composite coatings (Saucy and Storme 2013; Katnov et al. 2012).

By selecting fillers, it is possible to significantly improve such characteristics of paints and varnishes as adhesion, moisture resistance, viscosity, pouring, weather resistance, mechanical strength, hardness, etc. (Rosenfeld et al. 1987). Size and shape of filler particles have direct impact on properties of paints and varnishes and coatings based on them (Kuznetsova et al. 2018). Carbonates and silicates are most widely used in the formulations of paints and varnishes. This review examines the most interesting works related to use of carbonate and silicate fillers, as well as their modification (Stroganov and Amelchenko, 2013).

Among natural materials, diatomite, or diatomaceous earth, is of particular interest as a filler in paints and varnishes. Diatomite deposits are found in all parts of the world. Diatomite is a light, soft, light-colored sedimentary rock formed mainly from the silicon microshells of single-celled algae—diatoms, of a wide variety of shapes and sizes, usually 10–200 nm in diameter. The main component of siliceous shells are amorphous silicic hydrates with varying water content (opals) (SiO^2–nH_2O). Diatomite has a large internal surface and contains up to 80–90% voids. This material has a mainly macroporous structure, pores with a radius of 4–40 μm make up about 15% of the total pore volume, and its bulk density is ~30 g/dm^3 (Nizina 2016).

It is known that high-quality diatomites containing 60% or more silica have many useful properties—low bulk density, heat resistance, porosity, low thermal conductivity, and sound conductivity. They are used as additives in cement, textile, petrochemical and food industries—as adsorbents and filters.

Significant share of diatomite is used for production of filter powders and is used as fillers in production of paper, plastics, and paints, in production of polishing materials and insecticides.

There is a known method for producing base of composite anti-corrosion paint and varnish material against rust, in which mechanical activation and dispersion of the composition with filler pigments is carried out in a chamber of an apparatus with a vortex layer of ferromagnetic particles. Processing resulting material in a chamber of apparatus with a vortex layer of ferromagnetic particles significantly increases rate of chemical reaction, activation of substance particles due to deformation of crystal lattice of macromolecules of material and sharp increase in chemical activity and degree of material dissociation.

The closest in technical essence and achieved result to proposed invention is a water-dispersion paint and varnish composition, protected by a patent (Zhdanov 2016), taken as the closest analogue (prototype).

Prototype composition includes aqueous acrylic dispersion, titanium dioxide. Prototype composition includes aqueous acrylic dispersion, titanium dioxide, kaolin (ultrasound crushed) and functional additives: surfactant, white spirit as antifoam, monoethylene glycol as a coalescing additive, sulfazim as thickener, ammonium salt of acrylic copolymer as a dispersant.

Creation of a water-dispersion paint and varnish composition is aimed at obtaining it without use of bead mills, without ultrasound, and does not require some of functional additives required in traditional (mentioned above) methods of their production.

Technical result from use of proposed invention is to simplify and speed up paint manufacturing process, reduce rate of its delamination, as well as reduce amount of expensive pigment in its composition while maintaining decorative and protective properties.

The Republic has large reserves of diatomite in the Aktobe region, and this allows it to be used as a filler in paint and varnish coating compositions.

Diatomite, mineral filler that prevents agglomeration of titanium particles, when modified, will be microsilica with silicon dioxide particles with developed pore structure at nanoscale level.

Studied water-dispersion paint and varnish composition based on modified diatomaceous earth will simplify and speed up paint manufacturing process, reduce delamination, reduce amount of expensive pigment in composition, and improve protective properties.

5.3 *Determination of Technology for Producing Polymer-Based Composites, Selection of Basic Equipment and Technological Scheme Development*

To resolve technical issues related to design of paint and varnish production (shops and areas for synthesis of film-forming substances, production of varnishes and paints and varnishes, painting production), issues of organization of work and, above all, questions about methods for carrying out this work must be carefully considered.

The main factors determining the method of organizing paint and varnish production are the quantity and nature of the products produced.

Organization of synthesis of film-forming substances and production of main types of paints and varnishes is based on horizontal scheme of movement of materials and processes (Fig. 1).

Paint and varnish production produces four types of finished products:

- transparent or colored varnishes (not containing pigment);
- paints and enamels;
- powder paints;
- special paints.

Solid raw materials (pigments and fillers) are stored on pallets in bags, barrels or containers. Solvents and solutions of film-forming substances are supplied from

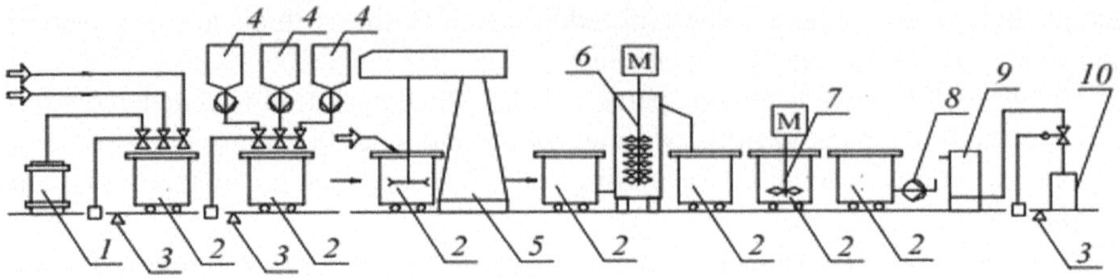

Fig. 1 Diagram of organization of production flow for manufacture of paints and varnishes, based on horizontal flow diagram of materials and processes: 1 – container for film-forming substance; 2 – bowl; 3 – scales; 4 – containers for tinting pastes; 5 – stationary mixer with a lifting stirrer; 6 – bead mill; 7 – mixer; 8 – pump; 9 – filter; 10 – packaging device; M – stirrer motor

storage tanks to pre-mixers or mixers with disc mixers through pipes through dosing devices.

Many correction tanks of different sizes are used. Mixed pastes are measured from a common paste mixer into mobile containers and then added to correction tanks. Preparation of mixture of paste or dispersion is carried out in adjustment tank by adding solutions of film-forming agents, solvents, and additives. Composition is monitored and adjusted (color, viscosity, other technically important properties).

Final product is sifted (powder paint) or filtered (liquid paints and varnishes), and drained from adjustment tank through filling line into containers or containers.

1) Functional responsibilities of the equipment:

Film-forming agent container: The initial stage of the process where the main component of the paint material is stored;
Bowl: The place where the main ingredients are mixed to create the initial mixture;
Scales: Used for precise dosing of components before adding them to bowl;
Containers for tinting pastes: Here are various tinting pastes to obtain the desired shades;
Stationary mixer with lifting stirrer: Used for homogenizing and mixing components;
Bead Mill: Place where ingredients are ground to achieve a specific texture or consistency;
Mixer: Additional means for final mixing and preparation of composition;
Pump: Used to pump the mixture along the production line;
Filter: Helps remove impurities and particles from the mixture before the final stage;
Packaging device: For packaging finished products before being sent to a warehouse or for sale;
M – stirrer motor: Provides movement of stirrer in stationary mixer for effective mixing of components.

2) Cyclogram of manufacturing process of paints and varnishes:

Preparation of components 5 min:

- Loading the film-forming substance into the container.
- Preparation of tinting pastes and other components.

Mixing and coloring 10 min:

- Dosing of the main components using scales and adding them to the bowl.
- Adding necessary tinting pastes to obtain desired shade.
- Mix using stationary mixer with a lifting stirrer until a homogeneous mass is obtained.

Grinding and additional mixing 7 min:

- Transfer mixture to a bead mill to grind the components (if necessary).
- Transfer mixture to additional mixer for final preparation of composition.

Filtration and packaging 8 min:

- Pumping mixture through a filter to remove impurities and particles.
- Packaging of finished products using packaging device.

Quality control 5 min:
Carrying out quality control at various stages of production process (possibly including analysis of viscosity, color, structure, etc.).
Final check and shipment 5 min:

- Final quality check before packaging.
- Shipment of packaged products to warehouse for storage or directly to points of sale.

Layout solutions for paint and varnish production must comply with scheme for transporting raw materials, product blanks and finished products adopted in technological process.

During designing, total area of workshop should be considered sum of production and auxiliary areas (without service and household areas).

Main indicators characterizing the specific size of the workshop area per unit of equipment:

- size of the total area;
- size of production area.

Size of production area per unit of main technological equipment should be determined:

- area occupied by a specific piece of equipment (passport data);
- width of passages and driveways;
- methods of loading raw materials into technological equipment, products onto conveyors of automatic lines, selecting finished products and removing finished products from conveyors by means of mechanization and automation.

Technological layout of workshop should include the most rational placement of all technological equipment and inventory, ensuring greatest convenience in work, the best transport connection between workplaces and production equipment and the smallest intra-shop cargo flow.

Special warehouses must be created to store basic stocks of paints and varnishes.

In workshop warehouses, stock of paint and varnish materials is allowed, not exceeding shift needs.

All premises for storing paint and varnish materials must be built in accordance with the building design and construction standards established for this category of buildings.

Preparation of materials for work should be carried out in a specially adapted fire-resistant insulated room (paint preparation room). Such departments should be in an isolated room near the workshop wall with window openings and emergency exit.

Study of state of current trends in paints and varnishes indicates prospects of using water-dispersion compositions based on polymer. Modern technological solutions for their improvement, to organize reliable and long-term protection, are based on modifying aqueous polymer dispersions with multifunctional surfactants.

6 Conclusions on the Fourth Chapter

1. In accordance with the set objectives, the analysis of technogenic waste of mining enterprises of the Republic of Kazakhstan was carried out, the experience of their use was studied, the qualitative characteristics of the enrichment waste of a number of mining enterprises (JSC Kazakhmys, JSC Achpolimetal, JSC Kazzinc) were determined. The analysis of the qualitative characteristics of the considered technogenic raw materials showed the feasibility of using the enrichment waste of the Zhezkazgan deposit and the polymetallic ores of the Kentau enrichment plant for obtaining dry mortars.
2. Analysis of literary data showed that the issues of modifying the surface of finely dispersed mineral fillers of dry mixtures and the development of finishing compositions based on them remains a very urgent task today. It was proposed to heat treat diatomite at a temperature of $t = 650$ °C in order to increase its surface activity, and a decrease in energy costs for activation was established, which causes dehydration of clay impurities to metakaolinite, which has increased reactivity.
3. Regularities of hardening of lime-diatomite finishing compositions depending on the dry mix formulation have been established. It has been revealed that the use of thermally activated diatomite increases the compressive strength of the lime-diatomite finishing composition, promotes an increase in early structural strength and accelerates the hardening processes of lime-diatomite compositions.

4. Study of the state of modern trends in paints and varnishes indicates the prospects of using water-dispersion compositions on a polymer basis. Modern technological solutions for their improvement, in order to organize reliable and long-term protection, are based on the modification of aqueous dispersions of polymers with multifunctional surfactants. Creation of environmentally friendly polymer protective composite materials that are resistant to external factors is one of the priority areas of researchers. It is known that currently the most common and affordable way to solve this problem is the use of filled water-based acrylic compositions.
5. Technology for production of polymer-based composites has been determined. Organization of production flow for manufacture of paints and varnishes is based on horizontal flow pattern of materials and processes.

References

Altayeva ZN, Baiserikova D, Bek AA (2019) Ways to regulate the properties of finishing dry mixtures using diatomite in their composition. Collection of materials of the XVIII annual Republican scientific student conference, Almaty. (in Russian)

Altayeva ZN, Baiserikova D, Abdrakhmanova L (2020) Features of technological and rheological properties of finishing compositions based on lime-diatomite binder. Collection of conference materials: Satpayev readings 2020: "Innovative solutions to traditional problems: engineering and technology", KazNITU named after K. Satpayev. (in Russian)

Bazhenov YM, Denisov GA (2003) Technology of dry building mixtures. ASV. 95 p. (in Russian)

Bezborodov VA (1998) Dry mixtures in modern construction. NGASU, Novosibirsk. 94 p. (in Russian)

Brock T (2007) European Guide to Paints and Varnishes and Coatings [Text]. T. Brock, M. Groteklaus, P. Mischke. Paint-Media. 548 p. (in Russian)

Dauletov ND, Suleimanov ST, Estemesov ZA, Zhakipbekov SK (1985) Effect of carbon-containing waste on the hardening process of binders. Complex use of mineral raw materials 7:73–77. (in Russian)

Demyanova BC (2001) Effective dry mixtures based on local materials. In: Demyanova BC, Kalashnikov VN, Duboshina NM et al (eds) Textbook manual for universities, 2nd edn. Publishing House ASV. 208 p. (in Russian)

Ermilov PI (1987) Pigments and pigmented paints and varnishes. Chemistry, p 200. (in Russian)

Fedulov AA (1999) Feasibility study of the advantages of using dry building mixtures. Construct Mater 3:26–27. (in Russian)

Goncharov MA, Chernyshov EM (2013) Formation of hardening systems of composites based on technological raw materials. Construct Mater 5:60–65. (in Russian)

Gorshkov VS, Timashev VV, Saveleev VG (1991) Methods of physicochemical analysis of binders. Higher School, Moscow. 334 p. (in Russian)

Korneev VM, Zozulya PV (2010) Dry building mixes. Nedra, 320 p. (in Russian)

Goreglyad SY (2001) The use of modifying additives in the production of dry construction mixtures. Construct Mater 8:28–29. (in Russian)

KNAUF in Russia and the CIS (2003) Advertising magazine. RIA "Kvartal" 1: 4–18. (in Russian)

Katnov VE, Stepin SN, Katnova MRR, Grishin PV (2012) Bulletin of the Kazan Technological University. 15(7): 95–96. 6. 128 Vlasova IN, The right filler – the right paint. Paints and varnishes and their application. 6–27–29 p. (in Russian)

Kazakova EE (2003) Water-dispersion acrylic paints and varnishes for construction purposes [Text] E. E. Kazakova, O. N. Skorokhodova. – M.: Paint-Media, 136 p. (in Russian)
Khaletskaya K (2014) Environmentally-friendly architectural water-borne paint for outdoor application: twenty years of experience in Belarus and Lithuania. K. Khaletskaya, V. Khaletski, S. Svediene, A. Mazeikiene. The 9th International Conference "Environmental Engineering" [Electronic resource]: Selected papers, Vilnius, Lithuania, 22–23 May 2014. Vilnius Gediminas Technical University. Electronic data. (415 Mb). Vilnius, 2014. 1 electron. opt. disc (CD-ROM). (in Russian)
Kolesnikova IV, Estemesova AS, Orynbekov ES (2019) Dry building mixtures on the raw material base of the Republic of Kazakhstan. Electronic. II International conference BALTIMIX ASIA 2019, 04/29/2019. (in Russian)
Kozlov VV (2000) Dry building mixtures: textbook. Manual for universities. ASV. 96 p. (in Russian)
Kuznetsova VA, Shapovalov GG (2018) Development trends in the field of erosion-resistant coatings (review). Proc VIAM Electr Sci Tech J 11(71):Art. 09. https://doi.org/10.18577/2307-6046-2018-0-11-74-85. http://www.viam-works.ru. (in Russian)
Loganina VI (1999) Local building materials. PGASA, Penza. 110 p. (in Russian)
Loganina VI, Puchkov RY (2002) Dry mixes for finishing building walls. Plast Synth Prop Process Appl 8:38–39. (in Russian)
Meshkov PI (2000) Methods for optimizing the compositions of dry building mixtures. Construct Mater 5:12–15. (in Russian)
Nizina TA (2016) Analysis of the influence of diatomite firing on the thermal conductivity of liquid heat-insulating coatings. Acad Bull UralNIIproekt RAASN 2:86–89. (in Russian)
Pavlov EA (2012) Paint and varnish industry. 10: 20–23. (in Russian)
Patent of the Republic of Kazakhstan for Utility Model № 2776 dated 09.04.2018, Method for Obtaining a Mineral Additive for Concretes and Mortars, Mineral Additive, Concrete and Mortar with a Mineral Additive (Estemesova A.S. Eselbaeva A.G.) (in Russian)
Project BR21882292, "Integrated development of sustainable construction industries: innovative technologies, optimization of production, effective use of resources and creation of technological park" with the support of the Science Committee of the Ministry of Science and Higher Education of the Republic of Kazakhstan (in Russian)
Rosenfeld IL, Rutinshtein FI, Zhigalova KA (1987) Protection of metals from corrosion with paint and varnish coatings. Chemistry, Moscow. 224 p. (in Russian)
Sartbaev MK, Zhalgasuly NZ, Estemesov ZA (2017) Construction materials from technogenic mineral formations. LEM Publishing House, Almaty. 224 p. (in Russian)
Saucy D, Storme P (2013) Paint and varnish industry 3, 18–20. (in Russian)
Shcherbakova MS, Shutilin YF, Moiseev VV (2013) Influence of paint and varnish composition on coating properties. VSUET Bull 1:135–137. (in Russian)
Shestoperov SV (1991) Concrete quality control. M. 274 p. (in Russian)
Skorokhodova ON (2013) Paint and varnish industry. 4 14-15. (in Russian)
Sokov VN, Bazhenova SI, Petrov MA, Pepelyaeva AY (2019) Development of the component composition of environmentally friendly water-dispersion paint based on polyvinyl acetate dispersion. Bulletin of BSTU named after V.G. Shukhov 3:15–21. (in Russian)
Stroganov VF, Amelchenko MO (2013) Influence of mineral fillers with different geometric shapes of particles in water-dispersed acrylic protective coatings and sealants on the adhesion properties to concrete and metal surfaces. Coll. t. Int. sci.-tech. conf. "Modern achievements in the field of adhesives, sealants. Materials, raw materials, technologies. Dzerzhinsk. pp 165–167. (in Russian)
Suleimanov ST, Estemesov ZA, Dauletov ND (1987) Effect of alkaline earth metal carbonates on the hardening process of slag-alkaline binders. Bull Acad Sci Republic of Kazakhstan Chem Ser 4:66–70. (in Russian)
Vasilik PG (2002) The use of fibers in dry construction mixtures. Construct Mater 9:26–27

Viktorovich AM (2000) DOW Chemical products for the building materials industry. Build Mater 5:10–12. (in Russian)

Yakovlev AD (2008) Chemistry and technology of paints and varnishes [Text]. In: Yakovlev AD (ed) Textbook for universities. KHIMIZDAT, St. Petersburg. 448 p. (in Russian)

Yestemesova AS (2019) Technology for the production of dry construction mixtures using broken glass. Monograph IOC (KazGASA) Almaty. 109 p. (in Russian)

Yestemesova AS, Yeselbaeva AG (2019) Modern energy-efficient building materials, products and technologies for their production. Tutorial IOC (KazGASU). 133 p. (in Russian)

Zharko VI, Guzov VA (2011) Raw material base of secondary resources in production of building materials. Int Analy Rev Cement Concr Dry Mixt (2):11–27. (in Russian)

Zhaparov AG, Vikul AY et al (2013) Some aspects of the production of construction materials in the process of developing technogenic deposits of Krivbaza. Construct Mater (7):66–70. (in Russian)

Zhdanov NN (2016) Influence of functionalized monomers on the properties of acryl-styrene dispersion and heat-protective coating based on it. Bull Technol Univ 19(23):65–67. (in Russian)

Study of Ash and Slag Waste for Their Use as Secondary Resources

Abstract This chapter presents the results of an environmental impact assessment of ash dumps resulting from the combustion of coal from the Ekibastuz deposit in Kazakhstan. It is shown that significant volumes of ash and slag waste have accumulated at the Ekibastuz GRES-2 ash storage facility and in the Almaty region due to the operation of three thermal power plants. These waste deposits pose serious environmental challenges.

The value of ash and slag waste is too high for it to be considered mere refuse. This study explores the potential for converting this waste into widely used construction materials. The urgency and significance of the issue are underscored by the fact that technogenic waste from thermal power plants in Kazakhstan is insufficiently processed. As a result, large volumes of ash continue to accumulate, occupying extensive land areas and effectively removing them from potential economic use.

Utilization of ash dumps offers a way to reduce anthropogenic impact on the environment while promoting the rational use of secondary raw materials. The chemical and granulometric compositions of ash and slag were investigated, along with X-ray phase and differential thermal analyses. The chemical profile of Ekibastuz ash provides insight into the mineral composition of the source coal, supporting its application in the production of construction materials.

The findings demonstrate feasibility of using ash and slag waste as a secondary raw material to reduce environmental impact. In particular, these materials have found application in the cement industry as slow-hardening binders, in road construction as active hydraulic additives combined with inorganic, bituminous, or polymer-bitumen binders, and as fillers or low-activity additives that can partially replace cement. Their use in the production of sand-lime bricks allows for significant savings in lime (10–50%, depending on the composition) and sand (up to 30%), while the final bricks are on average 10% lighter than conventional ones.

In ceramic production, ash and slag serve as burnout and fluxing additives, and in ash ceramics, they can act as the primary raw material. The largest volume of ash and slag waste is used in road construction—for leveling uneven terrain and as a filler in asphalt concrete pavements.

M. Nurpeissova et al., *Development of Technology for Processing of MMC Waste*, SpringerBriefs in Earth Sciences, https://doi.org/10.1007/978-3-032-12976-5_5

This chapter also outlines innovative methods developed by the authors for the utilization of ash and slag waste, including:

- Production of ash-containing binders;
- Fabrication of gold-containing ceramic products;
- Synthesis of algopoirite—a filler for lightweight concrete;
- Production of non-autoclaved ash concrete;
- Development of bitumen-ash-based binders for road surfaces;
- Composition and method for preparing splitter blocks;
- Composition and method for producing geopolymer ash concrete;
- Composition and method for self-compacting ash concrete;
- Composition and method for manufacturing ash fiber cement panels for facades.

1 Assessment of Ash and Slag Waste as a Source of Environmental Pollution and as a Source of Secondary Raw Materials

Kazakhstan has a significant number of thermal power plants. Every year, the amount of ash and slag waste (ASW) generated at thermal power plants (TPP), state district power plants (SDPP), as well as in boiler houses increases. Fuel and electricity complex is one of the main "pollutants" of the natural environment. By burning coal, enterprises obtain thermal energy and generate electrical energy. Negative side of this process is formation of by-products of coal combustion—fly ash (fly ash) and slag.

Currently, more than 500 million tons of ash and slag waste have been accumulated in Kazakhstan, which will increase to 1 billion tons by 2030. The lion's share of these technogenic materials is formed during combustion of Ekibastuz coals, which are hard, low-caking, high-ash, low-grained, but cheap. Therefore, they are used as fuel in many thermal power plants in Kazakhstan and Russia.

Deterioration of environmental situation is not unreasonably associated with air pollution. Long-term storage of thermal power waste in ash dumps allows harmful substances and heavy metal ions to enter water and soil. Anthropogenic component of formation of water surface quality is already comparable with natural component, which poses threat to sustainable water use. Annual yield of ash and ash and slag mixtures during burning coal in ash dumps in Kazakhstan is more than 17 million tons. More than 300 million tons of ash waste have been accumulated in ash dumps.

During combustion of Ekibastuz coals in boilers of thermal power plants, complex chemical and phase transformations of their mineral components occur as follows (Nurpeissova et al. 2023a, b):

- high-temperature processes associated with clay:

- at temperatures 450…650 °C kaolinite (Al_2O_3 $2SiO_2$ $2H_2O$), transforms into chemically active anhydrous metakaolinite (Al_2O_3 $2SiO_2$);
- at temperatures above 900 °C mullite ($3Al_2O_3 \cdot 2SiO_2$) is formed;

– processes associated with decomposition of carbonates:

- at 400…500 °C iron carbonate ($FeCO_3$) decomposes;
- at 600…1000 °C, dicarbonization of calcite ($CaCO_3$), magnesite ($MgCO_3$) and dolomite ($CaCO_3$ $MgCO_3$) occurs
- at 400…700 °C oxidation of pyrite (FeS_2) occurs with the formation of iron oxides (Fe_2O_3) and sulfur (SO_3);
- at high temperatures (1200 °C) eutectic mixtures appear, which are a glass phase, including a microsphere.

The following can be added:

- now there are more than 300 technologies for processing ash and slag waste, depending on their chemical and phase composition;
- the most ash-and-slag-intensive technology is their use in construction, including road construction.

Table 1 shows the negative consequences that arise from the impact of an ash dump on the ecosystem that these man-made wastes have on the biogeocenosis.

Table 1 Negative consequences arising from impact of ash dump on ecosystem

Ash dump	Buildings and structures	Conditions reduction for archophytomeliorative measures
		Degree reducing of biopositivity of buildings and structures
	Biocenosis [flora (F_2), fauna (F_n), human (H_s)]	Reduction strength characteristics and durability of buildings and structures
		Oppression and deterioration of human, flora and fauna
		Occurrence of various diseases
	L—Lithosphere	Alienation of a vast territory and disruption of the natural landscape
		Oppression of soil biota (earthworms, woodlice, mites, nematodes) and flora (fungi, bacteria, algae, etc.)
		Violation of functional properties of soils (turf, forest litter, meadow felt, humus horizon)
	A—Atmosphere	Increased ambient temperatures and climate change
		Formation of technogenic systems in the form of acid rain, the greenhouse effect, ozone holes, etc.
		Changes in natural functional properties of the air basin
	G- Hydrosphere	Anthropogenic impact of precipitation in the form of acid rain
		Increase in the amount of polluted ground and surface water
		Changes in the functional natural, physicochemical properties of aquatic systems

The lithosphere experiences the greatest anthropogenic load from the impact of technogenic waste among ecotope system. Characteristic of anthropogenic impact of components of ash dumps on lithosphere system is:

- suppression of biochemical processes occurring in soil system as a result of occurrence of accompanying silicate-forming physical and chemical processes;
- change in the pH of the reaction medium (pH of the soil medium is neutral, i.e. is about 7), both downward (with acidic reaction) and upward (with alkaline reaction), which leads to disruption of the redox reaction, occurring in the soil system;
- formation of various soluble and insoluble salts that inhibit soil components and negatively affect biochemical soil processes;

Emergence of centers of cementation through the formation of aluminosilicate systems on the surface of the earth, as a result of which vegetation is destroyed.

All this ultimately leads to erosion, pollution, salinization, desertion, dust, alienation and other negative consequences in the lithosphere.

By and large, ash is not a waste, but a valuable raw material—technogenic mineral formation. Therefore, in England and Germany, ash and slag are completely used in the national economy (100%), in Japan—82%, in the USA and China—65%, in Russia—15%.

However, for various reasons, ash dumps are not used in Kazakhstan, since the physical and chemical properties are very poorly studied, in addition, when they are processed, other waste is obtained from these wastes or ash dumps are unsuitable for use, because they contain a lot of unburnt coal. There are factors of disinterest. Therefore, ash dumps in storage facilities across Kazakhstan are increasing from year to year, causing, as mentioned above, environmental, economic and socio-economic damage amounting to a billion tenge.

- world experience shows that fly ash can be used in various areas of the national economy;
- however, the most ash-intensive and environmentally-economically efficient areas, as already mentioned, are the construction and road industries.

During using fly ash as mineral additive for cements and concrete mixtures, their physical, mechanical and operational properties are significantly improved. In connection with the above, relevance of using fly ash as component for cement and concrete mixtures is beyond doubt.

Purpose of this work is to study physicochemical characteristics of fly ash at Ekibastuz State District Power Plant-2, which is the most powerful thermal power station in Kazakhstan, in terms of its formation and accumulation. This will make it possible to use it more rationally and efficiently in the production of CEM II/A-Z, CEM II/V-Z, CEM V/A and CEM V/B.

Relevance and significance of this problem is enhanced by the fact that it is recognized that technogenic waste from thermal power plants is not processed, current ash waste accumulates and occupies huge areas (about 8.5 thousand hectares are used for storage), which removes it from land use (Fig. 1).

Fig. 1 Warehousing and storage of ash and slag waste

Development of electricity production and recycling of waste from thermal power plants, in particular ash from coal combustion, are considered one of the main government priorities of Kazakhstan.

Ash dumps are the reason for alienation of large areas of land for purpose of constructing ash dumps for placement of ash waste, which are almost irretrievably withdrawn from useful use, even after their reclamation, and their maintenance requires significant operating costs, which increases the cost of energy production. One of the problems with storing ash waste at ash dumps is their location near large cities (and often within city limits). Problem of surface deformation and relief changes arises.

In this work, we studied the ash and slag waste of CHPP-2 and CHPP-3 of the Almaty power plant, which combines 3 thermal power plants that provide heat and electricity to consumers in Almaty city and the Almaty region of Kazakhstan. All thermal power plants use coal from the Ekibastuz deposit. Ecological situation in Almaty is complex. The Development Strategy "Almaty-2050" reports that more than 122 thousand tons of harmful substances are emitted into the air every year in the city. Sixty-five percent comes from transport. This is more than 500 thousand city cars and about 200 thousand non-resident cars. Twenty-seven percent of emissions, according to data of local administration come from CHPP-2, which burns high-ash Ekibastuz coal. The ash dump of Almaty CHPP-2 is located within the Alatau district of the city (Fig. 2).At the ash dump of CHPP-3. ordinary samples weighing from 3–5 to 15–16 kg were selected, from which group samples were subsequently compiled (Fig. 3).

Methodology XRF and RFSA were used to identify chemical and phase compositions, and standard methods were used to determine the physical properties.

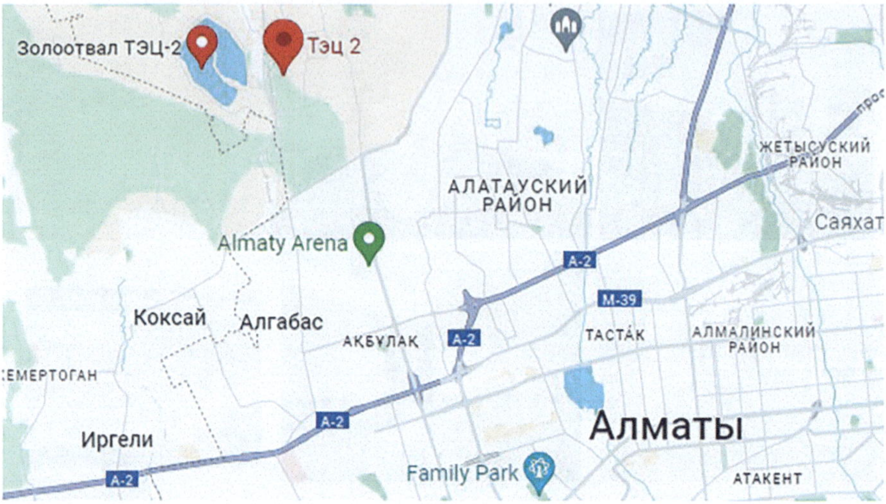

Fig. 2 Location of the ash dump TPP-2 in Almaty

Fig. 3 Sampling from ash and slag waste

To determine phase composition of cements and ash (to obtain strength), upgraded DRON-3 M diffractometer on CuKα radiation with software was used. X-ray patterns were obtained in the range of 2θ (angles) from 10 to 70°.

Knowledge of chemical composition of ash and slag waste is necessary condition for judging its properties and deciding on possibility of using it in various sectors of the national economy. To determine chemical composition of ash and slag waste, ash samples were taken from electric precipitators and an ash and slag dump according to the methodology RD 34.09.603–88.

The particle size distribution analysis was carried out using an Analizette 22 MicroTec Fritsch GmbH (Germany) device. Microphotograph of fly ash was taken using a Superprobe-733 scanning electron microscope with software.

Intensification in this direction involves use of industrial waste instead of primary natural resources to reduce cost of building materials. In this direction,

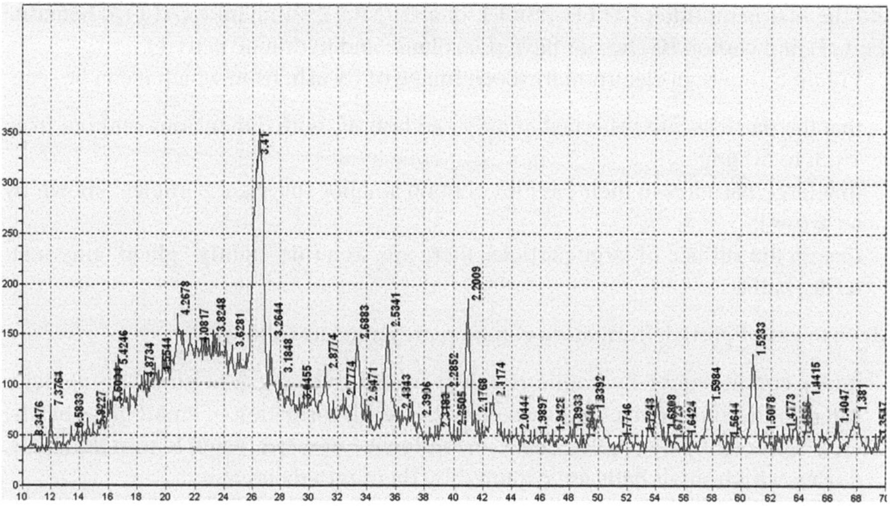

Fig. 4 X-ray image of fly ash

scientists at Satbayev University are conducting a large amount of research on production of secondary raw materials based on industrial waste.

Research Results By burning coal, thermal power plants receive thermal energy and generate electrical energy. Negative side of this process is formation of by-products of coal combustion—fly ash and slag. Composition of the ash and slag material was determined by quantitative ratio of its constituent minerals, which depend on the mineralogical composition of the initial part of fuel.

Chemical and Phase Compositions Figure 4 shows X-ray diffraction pattern of ash, from which it is clear that reflections with interplanar distances are recorded, d/n, Å:

- mullite ($3Al_2O_3{\cdot}2SiO_2$)—54,246; 3,41; 22,009 and others;
- quartz (SiO_2)—42,678; 3,41; 22,852 and others;
- sillimanite ($2Al_2O_3{\cdot}SiO_2$)—54,246; 3,41; 15,233 and others

At the same time, values of some mineral peaks coincide with each other, especially between mullite and sillimanite. In addition, ash also contains a glass phase, as evidenced by a halo on the X-ray diffraction pattern in the region of 16° … 28° and unburned carbon, which is not visible on the X-ray diffraction pattern.

Quantitatively, the ash components are distributed as follows, %: mullite—38; quartz—32; sillimanite—12; hematite—5; glass phase—10; unburned carbon—3.

Chemical composition, %: SiO_2–57,7; Al_2O_3–29,6; (Fe_2O_3+ FeO)—6,4; CaO—1,1; MgO—0,35; SO_3–1,3; K_2O—0,03; Na_2O—0,52; lop—3,0.

Analyzing phase composition of fly ash, it can be stated: from individual, only the glass phase, which contains microsphere, has pozzolanic and hydraulic activity, and the rest is mullite ($3Al_2O3\ 2SiO_2$), quartz (SiO_2), sillimanite (Al_2O_3), hematite (Fe^2O3) and carbon (C) do not have pozzolanic and hydraulic activity.

Figure 5 shows an electromicroscope image of fly ash, from which it can be seen:

- that the particles are spherical, glassy and hollow, with dimensions ranging from 1 μm to 50 μm;
- that large particles in their cavities contain smaller spherical particles (shown by an arrow);
- that on the surface of large particles there are, as a rule, tightly "glued" tiny scattering balls.

Mechanism of particle formation can be represented as follows:

- under conditions of hydraulic removal from furnace, ash at elevated temperatures comes into contact with water, resulting in formation of small glass beads;
- after a moment, large melts, due to the influence of water, begin to turn into balls, during which small balls are captured by them into their cavity;
- "massive balls" cool more slowly, so small balls that are already cooling stick to their surface;
- it is possible for relatively large balls to appear, inside of which there are small balls, according to the following mechanism: small balls pile up in one place, attracting each other, then in the process of rotational movement, the outer balls still stick tightly together, thereby creating a continuous outer surface; By the way: small balls located in the cavity of a large ball, as a rule, are mullite and α-quartz (Petrographic 2022).

It should be especially noted that on the microscope screen the balls and beads of fly ash have a shiny and white (light) surface, which is characteristic of the texture of

Fig. 5 Micrograph of ash fractions in a scanning electron microscope

glass. This color fades in the photo. Particles of unburned carbon here and there give a black background.

The granulometric composition of the Ekibastuz State District Power Plant is distributed among fractions as follows: up to 0,5 mm—0.14%; 0,45 mm—2,26%; 0,25 mm—3,6%; 0.1 mm—25,8%; 0.09 mm—0,84%; 0.08 mm—12,12%; 0,06 mm—4,5%; 0,05 mm—21,46%; 0.045 mm—21,38%; 0.04 mm—7,9%. At the same time, the percentage point is about 3%.

Table 2 shows distribution of phase composition of fly ash depending on its fractional composition. Analyzing the data in this table we can say the following:

- fly ash is mainly represented by a fraction consisting of particles with a size of 100 microns—25.8%, with a size of 80 microns—12.12%, with a size of 50 microns—21.46%, with a size of 45 microns—21.38%; in total they reach 80.76%;
- composition of a particle with a size of 500 microns stands out, which is represented by mullite (28%), α-quartz (54%), carbon (18%) and a small amount of glass phase (occupying a halo area on the X-ray diffraction pattern reaches 6 cm^2; their content of the total amount fly ash is small—only 0,14%;
- content of unburned carbon on X-ray diffraction patterns is recorded only in the composition of large fractions—from 500 μm to 63 μm; Moreover, within these fractions its content naturally decreases: 18%–7%—from coarse fraction to the fine;
- the halo area on the X-ray diffraction patterns (Fig. 6), representing the glass phase, increases with decreasing dispersion of fly ash; for example, the halo area on the X-ray diffraction pattern of particles 500 μm in size reaches 6 cm^2, its intensity is 105 pulses/sec, and the size of 40 μm is, respectively, 23 cm^2; therefore, in the latter composition of fly ash there is more (almost 3.8 more and 140 pulses/sec) glass phase, therefore, small particles may have the greatest activity;

Table 2 Distribution of phase composition of fly ash depending on its fraction

Sieve mesh number	Particle sizes, microns	Fraction content on the sieve, %	Distribution of phase composition depending on the fraction, %				
			Mullite	α-quartz	Silimannit	Carbon	Glass phase (halo area occupied, cm^2)
0,5	500	0,14	28	54	–	18	6
045	450	2,26	44	21	24	11	14
025	250	3,36					
01	100	25,8	46	18	28	8	14,5
009	90	0,84					
008	80	12,12	42	20	31	7	18
0063	63	4,5	47	17	29	7	20
005	50	21,46	50	21	29	–	22
0045	45	21,38	47	25	28	–	23
004	40	7,9	51	14	35	–	23

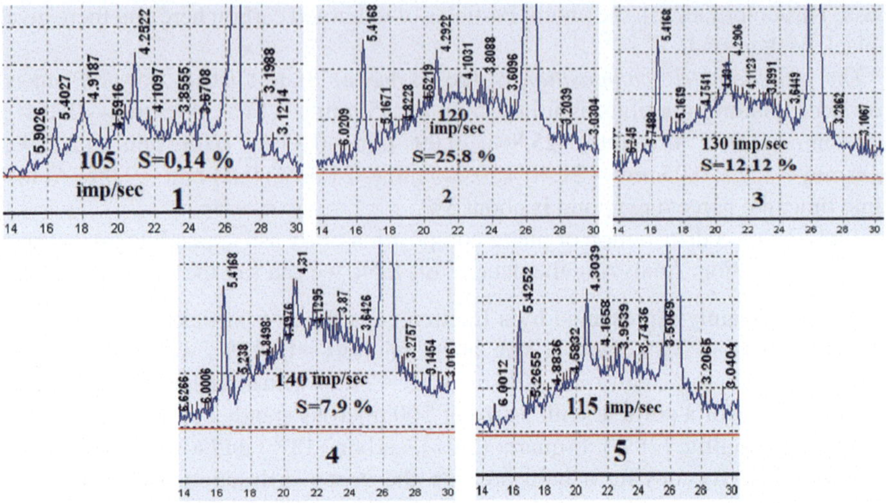

Fig. 6 —Distribution of glass phase in fly ash fractions: 1…4, respectively, fraction 500, 100, 80 and 40 μm; 5—glass phase content in the initial fly ash

- in fly ash particles with a size of less than 450 microns, the content of mullite ranges from 41…51%, α-quartz—14…25% and sillimanite –24…35%. It is known that the above-mentioned components of Ekibastuz fly ash have the following characteristics:
  - ***mullite and sillimanite*** are representatives of andalusite group; have hardness in the range of 6…7.5, density—3.03…3.23 g/cm^3; optical characteristics in transmitted light—light refraction of sillimanite—Ng = 1677; Nm = 1658; Np = 1657; mullite—Ng = 1654; Nm = 1644; Np = 1642; have high fire resistance, are chemically extremely inert and have increased mechanical strength;
  - ***α-quartz*** is a low-temperature form of the mineral; uniaxial, positive: Ng = 1553 and Np = 1544; density 2,65 g/cm^3; hardness—7; almost insoluble in H_2O, HCl and H_2SO_4, soluble in HF, weakly in alkalis. Moreover, it should be noted that 26% of α-quartz is reactive towards lime.
  - ***hematite***—density 5,26 g/cm^3, hardness 5…6, uniaxial, negative; Ng > 2.95, Ne = 2,74. Insoluble in water, organic acids, soluble in concentrated HCl solution;
  - ***coal***—is amorphous carbon, density in the range of 1.8…2.1 g/cm^3; consists of tiny crystals that are invisible even under a microscope and have graphite structure; has absolutely no astringent properties; decomposes from the sun's rays and does not create compounds with water; does not interact and has no bonding force with the minerals of Portland cement clinker, lime, gypsum, mineral components of crushed stone, gravel and sand, therefore its presence in composition of cement stone, mortar, concrete and reinforced concrete products and structures is highly undesirable

- ***glass phase***—occurs as a result of sudden cooling with water; has different densities: if it is lighter than water, then such a glass phase is called a microsphere, which has increased hydraulic and pozzolanic properties; therefore it is used as an active mineral supplement; if its density is higher than water, then it is known as a glass phase, which has latent hydraulic and pozzolanic properties.

Both types of glass phase actively interact (especially the microsphere) with the hydration products of Portland cement, forming additional cementing substances in hardening systems.

Thus, it can be stated that the fly ash fraction of Ekibastuz GRES-2 consists of three groups:

- **the first group**—it includes mullite, sillimanite, hematite and some part of α-quartz; they play the role of filler in composition of cement stone, mortar and concrete, which, under certain conditions, helps to increase the density and strength of the latter; their content in fly ash reaches 61% by weight;
- **the second group**—it includes the glass phase, the amount of which in fly ash reaches 10% and the reactive part of quartz (26%). The glass phase is capable of interacting with aqueous compounds (especially $Ca(OH)_2$) of hardening systems, supplying additional cementitious substances, which will have a very beneficial effect on the density, strength and durability of products and structures. Reactivity of these compounds decreases in the order: microsphere>glass phase>active quartz;
- **the third group**—this includes unburned coal, which is a harmful ballast in hardening and hardened binder systems, therefore its presence in the composition of cement stone, mortar and concrete is highly undesirable; nevertheless, it is always present in fly ash; use of concrete in its composition is carried out in accordance with the technical requirements of GOST 25818–2017.

According to the technical requirements of GOST 31108–2020, fly ash is introduced into the following types of cements (Panteleev et al. 1978):

- CEM II/A-Z in an amount of 6...20% by weight and CEM II/V-Z in an amount of 21...35%;
- CEM V/A in the amount of 18...30% and CEM V/B in the amount of 31...49%.

In this case, fly ash must meet the technical requirements of the above-mentioned standard. In Table 3 shows the chemical indicators of the standard and the studied fly ash.

Currently, more than 500 million tons of fly ash have been accumulated in Kazakhstan, which will increase to 1 billion tons by 2030. This accumulation of fly ash in storage facilities creates powerful anthropogenic impact on biogeocenosis. It is obvious that complete recycling with production of necessary materials and products based on it contributes to creation of low- and low-waste technologies and the rational use of natural resources, which leads to solving the environmental problem of the environment. In light of this, comprehensive physical and chemical studies of

Table 3 Comparative data of chemical indicators and fly ash

Name indicator	ND for testing	Normative documents standard according to GOST 31108-2020	Actual value fly ash
Alkaline oxide content (R_2O) in terms of Na_2O, no more, %	GOST 5382-2019, p.12	2,0	0,54
Content MgO, no more, %	GOST 5382-2019, p.7.3	5,0	0,35
Mass loss on piercing (lop), no more, %	GOST 5382-2019, p.4	5,0	3,0
Uniformity of volume change (expansion of cement with the addition of fly ash, no more than, mm	GOST 30744-2001, p.7	10	9,0

fly ash from the ash storage facility at Ekibastuz State District Power Plant-2 were carried out.

The problem of slag utilization in construction remains urgent task, since almost all research is limited to experimental developments. All this raises an urgent need to conduct targeted comprehensive studies of both the slag itself and materials based on it. Therefore, today waste management has become particularly relevant, as one of the key areas for the development of a "green" economy in Kazakhstan, i.e. conservation and effective management of ecosystems (Nurpeissova et al. 2022; Begentayev et al. 2024).

The growing scale of construction in Kazakhstan requires significant amount of mineral raw materials for the building materials industry. Intensification in this direction involves the use of industrial waste instead of primary natural resources in order to reduce the cost of building materials. The use of solid mining waste in the building materials industry is more economical compared to the production of building materials based on special mining of mineral raw materials.

2 Use of Ash and Slag Waste from Thermal Power Plants in Conditions of Greening and Introduction of Promising Technologies

Despite the rapid development of modern technologies in energy industry, coal still remains one of the main sources of energy throughout world. Of course, like any other type of energy resource, the fuel and electricity complex is one of the main "pollutants" of the natural environment. These are air emissions (48% of all air emissions), wastewater discharges (36% of all discharges), and solid waste generation (30% of all solid pollutants).

Deterioration of environmental situation is not unreasonably associated with air pollution. Long-term storage of thermal power waste in ash dumps allows harmful substances and heavy metal ions to enter the water and soil. Anthropogenic component of formation of water surface quality is already comparable with the natural component, which poses a threat to sustainable water use. Ash and slag waste is stored in ash dumps, which are essentially a simple pit created in the open air. Toxic and poisonous chemical compounds enter the soil, groundwater, and air and cause not only harm to the environment, but also directly to human health.

Problem of ash dumps in Kazakhstan requires early resolution in favor of environment, and not of individuals aimed at making profit from the ash and slag waste accumulated since the times of the Soviet Union. So, back in 2001, the Ministry of Emergency Situations of the Republic of Kazakhstan made a statement that "in the city of Astana, at CHPP-2 there are problems with storing ash and slag waste, which is located right in the city. The current Kyzylorda thermal power plant, which switched to gas and fuel oil, left behind a huge ash dump territory of which is already in contact with residential areas of the city (Zhalgasuly et al. 2019).

One of the largest thermal power plants is JSC "Almaty Electric Station" CHPP (AlPP). The AlPP CHPP complex is located 16.5 km from the center of Almaty and combines 3 combined heat and power plants (CHP-2, CHPP-3), which provide heat and electricity from consumers in Almaty and the Almaty region of Kazakhstan. All thermal power plants use coal from the Ekibastuz deposit (Nurpeissova et al. 2023a, b).

One of the most promising areas for use of waste from fuel and energy complex is their involvement in secondary circulation as secondary material. By involving industrial waste, it is possible to radically change the parameters of the Republic's raw material base.

Currently, KazNRTU named after K.I. Satpayev is building plant for processing industrial waste, including ash and slag waste. Plant will process ash coming from CHPP-2 and CHPP-3. The ash will be used to produce raw materials for production of construction industry products (Kuldeyev et al. 2022; Nurpeissova et al. 2023a, b).

There is no doubt that construction of plant for processing ash and slag waste must begin in close proximity to coal-burning stations. This will increase production profitability. In the scientific literature there are quite a lot of publications on the use of fuel and energy waste in various areas of construction industry. We analyzed scientific publications of domestic and foreign researchers in the field of development of technology for utilization and processing ash and slag waste from thermal power plants into useful products. The largest number of publications on the use of ash and slag waste relates to construction industry, including technologies development that allow use of waste in several directions. We will focus on some of them.

Cement Production A number of works have examined possibility of using ash and slag waste from thermal power plants (TPP) and hydroelectric power plants (HPP) in cement production with mineral additives.

It has been established that their physical and mechanical characteristics depend on composition of materials used, the amount of mineral additives and the specific surface area. The introduction of 15% acidic ASW into cement composition makes it possible to obtain good quality materials. In addition, fly ash is suitable for producing gypsum binders. Adding it to the raw material mixture in amount of up to 20% leads to increase in the strength and water resistance of the resulting compositions.

Concrete Production Thermal power waste can be used in production of various types of concrete. Concrete based on fly ash cement is not inferior in performance to concrete based on conventional cement, and exceeds it in water resistance. Promising filler for lightweight ash concrete is unfired ash gravel, obtained from high-quality fly ash and silicate blocks. Advantage of this material is that its thermal conductivity is reduced by more than 1,5 times compared to expanded clay concrete (Entin et al. 2012; Efremenko and Khaltaeva 2014).

Production of Ceramic Products To reduce consumption of energy and mineral resources in the production of ceramic materials, a promising solution is the use of ASW. Many researchers are working on this problem. The works of scientists show that ash and slag waste from thermal power plants can replace high-quality raw materials in the production of building ceramics (Kuldeev et al. 2021; Malchik et al. 2016). Effect of waste modification by grinding on physical and mechanical properties of the final product is shown using example of ash from Tomsk State District Power Plant-2. It has been established that increased content of fine particles of alumina has positive effect on water absorption and frost resistance of bricks produced by semi-dry pressing at a pressure of 25 MPa. Moreover, amount of waste ash additive in charge composition can be increased to 50%. Process of producing ceramic tiles can also involve waste from fuel and energy complex. It has been shown that by varying content of the main components, it is possible to change color and technical characteristics of resulting materials (Yuryev et al. 2013).

Production of Thermal Insulation Materials Ash and slag waste is also suitable for production of thermal insulation materials. The work (Abdrakhimov 2018) shows that lightweight brick can be obtained from a charge containing ash from the Tolyatti thermal power plant and intershale clay formed during the extraction of oil shale. Distinctive feature is that technological process does not contain traditional natural raw materials.

Road Construction Ash and slag waste from thermal power plants is a universal material for road construction. They can be used for filling road embankments, constructing foundations and all layers of highways, as a component of binding materials for strengthening soils. Use of ash and slag makes it possible to obtain complex cementitious materials, including metallurgical slag and Portland cement, which serve to strengthen foundations of highways and make it possible to reduce the cost of laying the roadbed by almost half (Gritsuk et al. 2016; Madanbekov and

Osmonova 2016). In terms of their physical and mechanical characteristics, fly ash asphalt concrete meets the requirements of state standards, and in terms of strength they exceed standard samples.

3 Development of Method for Obtaining Ash-Containing Binder

Problem of rational use of cement in building materials and products can be solved through development and use of binders, as well as improvement of technological methods that ensure required performance properties of products based on them. In Kazakhstan and abroad, effective low-cement compositions have been developed using technogenic waste, including tailings from processing plants, fly ash from thermal power plants, etc.

Invention relates to the field of processing ash and slag waste from coal thermal power plants for the purpose of their disposal as fillers in the production of construction products, as well as to ensure, if necessary, possibility of dust-free storage in dumps. Technical problem is to simplify technology for obtaining binder, reducing number of ingredients while simultaneously increasing its strength.

Technical result when using proposed binder is method for producing high-strength ash-cement binder with simplified technology, fewer components with increased strength and using unit capable of carrying out mechanochemical activation of the ash-Portland cement mixture to the greatest extent.

There is extensive patent and scientific and technical information on ash and slag utilization, which can be grouped into three areas:

- hydraulic removal of the entire mass of ash and slag with its subsequent disposal,
- dry removal of the entire mass of ash with its subsequent disposal,
- partial utilization of ash mass in the production of building materials.

The purpose of the work was to develop composition and study strength properties of a binder based on fly ash from the Ekibastuz State District Power Plant.

There is a known method for producing a binder (Patent RU 2 452 703, 2012). The following components were used:

- acid fly ash from thermal power plants;
- bauxite or nepheline sludge;
- Portland cement;
- complex additive consisting of gypsum (G-2 GOST 125-79), superplasticizer (TU 5745-022-58042865-2007), calcium chloride (GOST 450-77).

Binder is obtained by joint grinding in devices of disintegrator operating principle at linear rotor speeds by less.

The disadvantages of this binder are:

1. multi-component (six components), which entails mandatory creation of six technological lines (this is only for binders; when using them in practice, at least three more technological lines are needed for water, crushed stone and sand—to obtain concrete products);
2. it is doubtful (to say the least) presence in the composition of the complex additive of semi-aqueous building gypsum—bassanite ($CaSO_4 \cdot 1.5H_2O$), since during the preparation of products it creates a false setting as a result of its transition to dihydrous gypsum according to the equation:

$$CaSO_4 \cdot 0{,}5H_2O + H_2O \rightarrow CaSO_4 \cdot 2H_2O$$

To regulate setting time and intensity of cement hardening when grinding Portland cement clinker, gypsum with the formula $CaSO_4 \cdot 2H_2O$ is also introduced into the mill. At the same time, operator ensures that temperature of the inside of the mill, where the cement is produced, does not exceed 140 °C. If the temperature of the cement exceeds this temperature, the dihydrate gypsum will dehydrate to form the above-mentioned bassanite, which will lead to a violation of the setting time and "false fast" of cement, which is unacceptable (Butt et al. 1980).

Moreover, according to the standard, $CaSO_4 \cdot 2H_2O$ (gypsum) is already present in the composition of the used Portland cement in sufficient quantity.

Use of calcium chloride in binders, which very actively corrodes reinforcement in reinforced concrete structures, so its use in modern building products is very limited.

All these disadvantages taken together (you can also add to them increase in the cost of technical lines) cause the practical value of this binder, i.e. limit its use.

1. There is another known method for obtaining binders (Patent RU 2 664 567, 2018). The components were used:
   - fly ash;
   - burnt rock;
   - drywall waste;
   - Portland cement clinker;
   - alkali-containing waste based on sodium hydroxide and sodium chloride in the ratio NaOH: NaCl = 1:1,5.
   - alkali-containing waste based on sodium hydroxide and sodium chloride in the ratio NaOH: NaCl = 1:1,5.

Depending on the ratio of the components, the strength of the hardened binder in compression after steaming reaches 29,65…41,15 MPa, in bending—5,09…6,41 MPa, after normal hardening in water for 28 days respectively: 42,35…54,87 MPa and 4.23…8,54 MPa.

Technical results of known method are achieved by grinding them, which consists of two stages:

- first stage—preliminary activation (grinding) of fly ash, burnt rock and/or their mixture is carried out by co-grinding with solid alkali-containing waste based on

sodium hydroxide and sodium chloride in the ratio NaOH: NaCl = 1:1,5 and in an amount of 5… 10% by weight of active mineral supplement;
- second stage—mixture from the first stage is mixed in specified proportions with Portland cement clinker and drywall waste, and grinding is carried out again.

Disadvantages of this binder are:

1. Multicomponent (5 components), this means: to use this binder at a concrete mixing unit, it is necessary to create 5 independent technological lines, in addition, when using this binder to produce concrete, at least three more technological lines are required for crushed stone, sand and water; therefore, there are 8 (eight) technological lines; this is too much, in practice it is difficult to implement, but the cost of technical lines will increase significantly.
2. The two-stage technological line for producing this binder itself, which further complicates its production.
3. Presence of harmful sodium chloride (NaCl) in the binder makes it attractive to use in concrete structures, since NaCl corrodes not only the reinforcement in them, but also the metal elements of production lines.
4. The outer part of the drywall is cardboard lining; This is paper that is absolutely incompatible with concrete mixture.
5. These disadvantages limit its use in production.

There is a known method for producing cement (patent RU 2 497767 C1, 2013; Patent RU 2 497 767, 2006). Qualitative and quantitative components content, compositions and strength are given in Table 4 (Zotkin 2014).

Disadvantages of this method of producing cement are:

- multicomponent;
- need to combine mineral additives during technological processes;
- relatively low strength of cement, since such strength (M 400 D0 or CEM I 32.5) can be obtained without complicating cement production technology.

The closest in technical essence and achieved result is method of producing binders consisting of cement and fly ash from the Ekibastuz State District Power Plant, by joint grinding in a ball or jet mill (Suleimenov 1996).

The obtained data are presented in Table 5, from which it is clear that:

Table 4 Qualitative and quantitative content of components (Zotkin 2014)

Components and strength	Composition%						
	1	2	3	4	5	6	7
Portland cement clinker	66,8	66,8	66,8	66,8	66,8	66,8	66,8
Gypsum stone	5,2	5,2	5,2	5,2	5,2	5,2	5,2
Fly ash	18	12,5	9	5,5	0	0	0
Compressive strength, MPa:							
- steamed at age 1 day	11,3	20,6	22,2	24,2	28,9	27,05	23,6
- hardening in water, age 28 days	40,2	40,4	41,9	43,6	45,2	44,1	39,2

Table 5 Method for obtaining binders (Suleimenov 1996)

Binder composition, wt.%		Specific surface density, cm^2/g	Bulk density, kg/m^3	Tensile strength, MPa, after steaming			
				When compressed		During bending	
Cement	Ash			1 day	28 days	1 day	28 days
100	without additional grinding	2980	1250	29,2	43,5	4,3	5,7
70	30	5730	980	49,0	58,0	6,3	7,0
50	50	5800	910	42,0	48,0	8,1	10,5
30	70	5860	865	33,0	37,0	7,1	9,5

- with final grinding in a ball mill and an ash content of 30…50%, the activity of the cement-ash binder after 28 days. and after steaming according to the mode 2 + 10 + 2 hours at an isothermal heating temperature of 95 °C was 45…58 MPa, exceeding the activity of the original cement (44 MPa);
- the activity of cement-ash binder with 70% ash has reached 33…35 MPa;
- the specific surface area of the original cement was 2980 cm^2/g, and that of the cement-ash cement was 5730…5860 cm^2/g;
- bulk density of original cement—1250 kg/m^3, and cement-ash with 30% ash—980 kg/m3, with 50% ash—910 kg/m^3 and with 70% ash—865 kg/m^3.

Thus:

- ground binders acquire maximum strength with the addition of 30% fly ash; subsequently, increase in fly ash content leads to decrease in their strength; but these indicators are higher than strength of pure cement;
- maximum increase in strength of steamed binders reaches 68% after 24 h, and at the age of 28 days—33%.

Disadvantage of known pre-milled binders is their relatively low strength. Technical result of claimed invention is to increase binder strength. The specified technical result is achieved by the fact that in the process of joint grinding of cement and ash in given ratios, dry superplasticizer NeoLit-400 0,3% by weight of cement is introduced into the ball mill with the following components, wt. %:

Portland cement 70–30 fly ash 70–30
superplasticizer NeoLit-400 (top) 0,3%

CEM I 32,5 was used as Portland cement, and fly ash from the Ekibastuz State District Power Plant, a dry superplasticizer based on polycarboxylate copolymers (TU 5745-004-379 419-84-2015), was used as a mineral component.

Fly ash is characterized by the following physical and chemical properties:

- specific surface—290 m^2/kg, true density—2,1 g/cm^3, bulk density—780 kg/m^3, specific effective activity of radionuclides—72 Bq/kg;

- chemical composition is mainly represented by SiO^2 (56.7%) and Al_2O_3 (28,6%), phase composition—mullite (38%), quartz (32%), sillimanite (12%), hemotite (5%), glass phase (10%) and coke (3%); the basicity module is 0,02, and the activity module is 0,5; super sour.

Pre-dried cement and fly ash in specified ratios are loaded into a ball mill. Grinding duration is 6 h. Moreover, after 3 h from the moment of loading cement and fly ash, the required amount of NeoLit-400 is additionally introduced into the mill. After unloading, the specific surface area is determined using the device.

From the resulting mixtures, samples measuring 4 × 4 × 16 cm are prepared and tested in accordance with GOST 30744-2001, consider the fact that superplasticizer is present in composition of cement-ash binders. Therefore, spread of cone was taken into account, which was kept within 110–112 cm.

Steaming of samples is carried out according to the regime of 2 + 10 + 2 h at an isothermal heating temperature of 95 °C.

The results obtained are shown in Table 6, from which it can be seen that:

- in comparison with the original cement without pre-grinding, the compressive strength of the pre-grind binder, consisting of cement and ash with the addition of superplasticizer NeoLit-400, is significantly superior, reaching 40,3...55,5 MPa (versus 26,2 MPa) after steaming and 65,5 ...75,8 MPa (versus 44,3 MPa) at the age of 28 days;
- the same, when bending: 5.8...7,8 MPa (versus 4.8 MPa) after steaming and 7.9...8.9 MPa (versus 45,3 MPa).

Table 7 presents comparative data on the compressive strength of the countertype and the proposed invention, from which it is evident that:

- with a binder composition of 70:30 (cement:ash), the strength of the stone made from it after steaming reaches 55.5 MPa against 49.0 MPa of the prototype (a difference of 6,5 MPa), the same—at the age of 28 days: 75,8 MPa against 58,0 (a difference of 17,8 MPa);

Table 6 Steaming samples at a heating temperature of 95 °C

Binder composition, mas. %		Specific surface area, cm^2/g	Water-cement ratio	Flow rate, mm Cone blur	Tensile strength, MPa			
					After steaming		At the age of 28 days	
cement	ash				When bending	When compressed	при изгибе	при сжатии
100*	–	289	0,45	110	4,8	26,2	5,3	44,3
70	30	550	0,33	112	7,8	55,5	8,9	75,8
50	50	570	0,35	112	6,7	50,3	8,3	70,3
30	70	580	0,37	112	5,8	40,3	7,9	65,5

Note* Original cement, without regrinding and superplasticizer

Table 7 Comparative data on the compressive strength of the prototype

Binder composition, %		Prototype data		Data of the alleged invention	
		Compressive strength (MPa) after steam curing		Compressive strength (MPa)	
Cement	ash	1 day under compression	28 days under compression	A day after steaming	At the age of 28 days.
70	30	49,0	58,0	55,5	75,8
50	50	42,0	48,0	50,3	70,3
30	70	33,0	37,0	40,3	65,5

- with a binder composition of 50:50 (cement:ash), the strength of the stone made from it after steaming reaches 50.3 MPa against 42.0 MPa (a difference of 8,3 MPa); the same—at the age of 28 days: 70,3 against 48 MPa (a difference of 22,3 MPa);
- with a binder composition of 30:70 (cement:ash), the strength of the stone made from it after steaming reaches 40,3 MPa versus 33,0 MPa (a difference of 7,3 MPa); the same at the age of 28 days: 65,5 MPa versus 37,0 MPa (a difference of 28 MPa).

Observed effects are explained by the following:

- during intensive grinding in the mill, the particles of cement and fly ash components are partially amorphized, which increases their reactivity;
- the particles of NeoLit-400 superplasticizer present in the mill are adsorbed on the active surfaces and centers, which contributes to finer grinding of the particles of the above-mentioned components;
- when the above-mentioned mixture is mixed with water, the effectiveness of the presence of N-400 increases even more, namely (Zotkin 2014)

  - due to its plasticizing properties, the required amount of water is reduced; despite this, the mobility of the mixture increases;
  - N-400 destroys floccules—aggregates of cement grains and fly ash, as a result of which the water trapped in them is released, thereby it will participate in hydration processes;
  - in the hardening hydration system (cement-ash-SP-water), the superplasticizer (SP) molecules split off the Na + ion, and the remaining anions consist of a hydrocarbon radical and a certain amount of hydrophilic SO3- groups distributed along the length of the molecule, as a result of which the anions are adsorbed "flat" on the positively charged particles of cement and fly ash and hydration products, transferring a negative charge to them, thereby increasing the adhesion of the particles of hydrated formations.

Thus, method has been developed for producing an ash-containing binder, including mixing and joint grinding in a ball mill of Portland cement and fly ash, characterized in that in order to increase its strength, it additionally contains NeoLit-400 superplasticizer with the following content of ingredients, wt. %:

Portland cement—70-30, fly ash—30-70, superplasticizer NeoLit-400 (above)—0.3%. Resulting mixture, consisting of cement and ash, is ground together in a mill at a speed of 80 rpm for 3 hours, and then the calculated amount of superplasticizer NeoLit-400 is added to this mixture, after which it is ground for another 6 hours, bringing the specific surface area to 550…580 m^2/kg. Technical novelty of the method for producing agloporite—a filler for lightweight concrete is confirmed by the Patent of the Republic of Kazakhstan (Patent No. 8579, 2023).

4 Development of Method for Obtaining Ash-Containing Ceramic Products

Purpose of these studies is to study physicochemical properties of the ash and slag waste from burning of coal of the Ekibastuz deposit of the Almaty City Power Plant, to determine possibility of obtaining in demand of demanded building materials. The relevance and significance of this problem of use as a source of the secondary resource to reduce anthropogenic load on environment, study of the possibility of obtaining ceramic materials with appropriation of ash and slag waste.

We studied the ash and slag waste of the Almaty power plant. The Almaty power plant combines 3 heat-power centers (TPP-2 and TPP-3), which provide the heat and electricity of consumers of Almaty and the Almaty region of Kazakhstan.

As a result of the interpretation of this radiograph, the following minerals were installed in the following quantities, in % of the crystalline phase: hematite Fe_2O_3 – 12,1%, quartz SiO_2 – 32,4%, sillimanitis Al_2SiO_5—25,9%, mullit $Al_{4,\ 95}Si_{1,\ 05}O_{9,\ 52}$—29,6%.

One of the main indicators of raw materials is their granulometric composition. The larger the particles content, the higher the material plasticity. Consequently, the raw materials will have high connectedness, which will positively affect strength characteristics of finished products, granulometric composition is also important for determining adsorption abilities of the material (Kuldeyeev et al. 2023).

According to results of granulometric analysis in the studied ash 60% of the particles is from 10 to 70 microns. Further work consisted in samples manufacture ceramic bricks with addition of ash and slag waste. In the work, method of obtaining laboratory ceramic bricks was chosen by method of plastic molding with various percentage of ash and slag waste annealing at various temperatures. For clarity, all data are reduced to Table 8 and Fig. 7. Figures show dependence graphs of temperature of firing and proportions.

Testing Samples Further work consisted samples in the manufacture of ceramic bricks with the addition of ash and slag waste.

After experiment on study of properties of ash and slag waste, the next step was the manufacture of ceramic samples in which this waste was included. For this, it was necessary to do several technological processes, including preparation of raw

Table 8 Physico -mechanical characteristics of ceramic bricks at various firing temperatures and different percentage of ash and slag waste

Content of ash in clay, %	Thermal conductivity, W/m °C				Water absorption, %				limit of strength at compression, kg/cm^2			
	Temperature, °C				Temperature, °C				Temperature, °C			
	700	800	900	1000	700	800	900	1000	700	800	900	1000
0	0,84	0,85	0,85	0,85	11	10	9	8	60	80	90	110
5	0,83	0,84	0,82	0,85	17	17	15	10	70	96	100	115
10	0,83	0,81	0,82	0,81	18	18	16	12	72	104	110	120
15	0,77	0,79	0,74	0,76	20	19	17	14	75	106	115	125
20	0,70	0,69	0,73	0,71	24	23	18	16	70	90	105	115
25	0,55	0,6	0,57	0,61	27	25	20	18	65	85	95	110

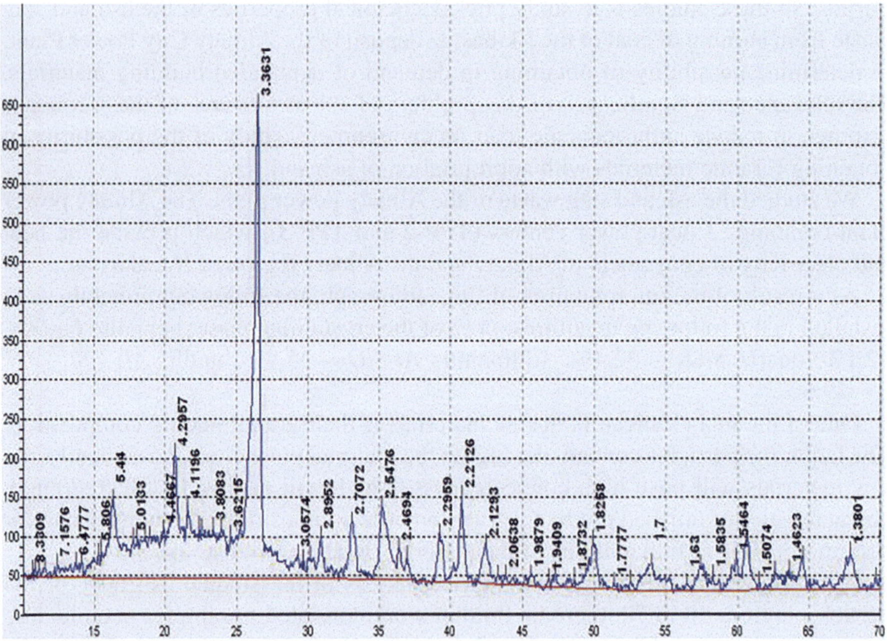

Fig. 7 Radiograph of the TPP-3 ash

materials, formation of ceramic samples and their firing in furnace at high temperature. After firing, series of tests was carried out to explore mechanical and physical properties of ceramic samples with ash and slag waste and compare them with samples without such an application. Tests results made it possible to conclude that streaming of ash and slag waste on properties of ceramic cowls on their strength and wear resistance.

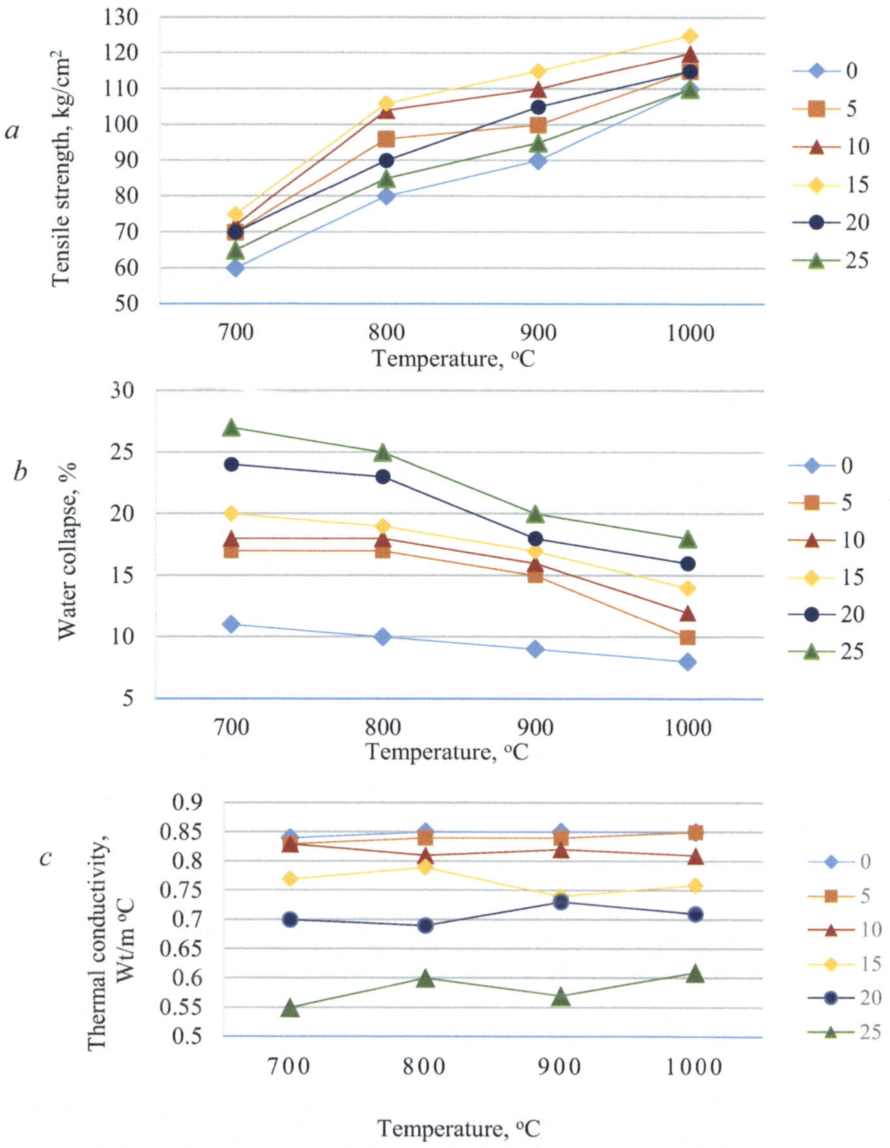

Fig. 8 Dependence graphs: a—tensile strength of the firing temperatures (0–25 Ash content in the clay); b—water collapse of the sample from temperature of the firing (0–25 of the ash content in the clay); c—thermal conductivity of the sample from the firing temperature (0–25 Ash content in the clay)

In the work, method of obtaining laboratory ceramic bricks was chosen by the method of plastic molding with various percentage of ash and slag waste and annealing at various temperatures. Figure 8 presents graphs of the temperature dependence on firing and ash content in the clay.

Analysis of the studies of ash and slag waste from burning of the Ekibastuz coals showed that thermal conductivity, strength, and water absorption depend on amount of ash and firing temperature added. The greater content of ash in the brick, less its thermal conductivity. Water absorption increased with increase in ash. Strength during compression also decreases with increase in content of ash in the brick. Optimal percentage of adding of ash and slag waste is 15% at a firing temperature of 1000° (Nurpeissova et al. 2024).

Based on studies results it is possible that future scientific studies will be aimed at optimizing production processes of ceramic bricks using ash and slag waste. It would be interesting to investigate influence of various parameters on brick properties, such as size of ash and slag waste particles, cooling rate and firing duration. In addition, further studies can be aimed at finding alternative materials for use in production of bricks, which can lead to development more stable and environmentally friendly building materials.

Analysis of chemical composition of Ekibastuz ash gives idea of composition of mineral substances of coal. Main components are silicon and aluminum oxides, a large amount of iron oxide is also present. It is necessary to know chemical composition of ash for solution about possibility of using it in various sectors of national economy.

One of the main indicators is the granulometric composition. The higher the content of microdispass particles, the higher the plasticity of the material, the product will also have greater strength and boundaries. Analysis of granulometric composition showed that in 60% of the particles the size is from 10 to 70 microns. Data shows that the material is very thin -dispersed.

Results of exploration showed possibility of using ash and slag waste as secondary raw materials to reduce anthropogenic load on environment.

5 Development of Method for Obtaining Agloporite—Lightweight Concrete Filler

Purpose of these studies is to study physicochemical properties of ash and slag waste from burning coal of the Ekibastuz deposit in the TPP-3 of Almaty city, to determine possibility of obtaining aggloporite-aggregate of the liccine concrete. Relevance and significance of this problem intensifies with fact that technological waste of thermal power plants is not sufficiently processed, current waste of ash accumulates and occupy huge areas, which removes them from land use. Utilization of ash damps allows you to reduce technogenic load on environment and ensure rational use of secondary raw materials.

Novelty lies in practical implementation of obtaining agglororite gravel of the "600" brand using ash of the TPP-3 in Almaty as a raw material that is ash of Ekibastuz coal. Study results make it possible to increase degree of complexity and environmental safety of development of technological waste-this is the development of industrial and innovative power of state of respect for natural resources and environment.

Practical *significance* lies in the fact that production of building products quickly pays off, does not require expensive installations, tools and equipment.

It is known that storage of ash and slag waste is very cost event. According to expert estimates, investments in reconstruction of one ash and slag waste damps can reach 5 billion tenge, and construction of a new one costs 10–12 billion tenge. Storage of ash and slag waste leads not only to removal of significant land areas, but also causes very significant pollution of almost all environmental components around their location (Ministry 2022). In the republic, annual output of ash, and particularly this coal, which belongs to stone gas fat coking, weakly spinning coals, has very high ash content, average of 42%.

Both the coal of the Ekibastuz deposit and its ash itself—to some extents are unique among the stone coals of other CIS deposits. In particular, this coal, belonging to stone gas coking, weakly spinning coals, has very high ash, on average of 42%.

Another distinctive feature of this coal is the fact that its ash is a very refractory. Such unusual stone coal for angles of the Ekibastuz ash is explained by features of the mineralogical composition of coal. According to source (Petrographic 2022), mineral impurities are quantitatively represented, in %: 54 kaolinitis, quartz—28, 10 siderite, calcite—5, gypsum—2 and magnesite.

It is not difficult to notice from this that the main minerals-intestes are very refractory minerals, such as kaolin (whose melting temperature is 1770 °C) and quartz (whose melting temperature is 1713–1728 °C). They even determine the very weak sintering and high melting temperature of the ash of the Ekibastuz coals, and the very low content of alloy minerals, such as: siderite, calcite, gypsum and magnesite, only partly reduce total temperature of sintering and melting of this ash. It is this mineralogical composition of impurities of the coal that translates ash of the Ekibastuz coals into category of unique in its refractory.

In addition, as a number of researchers notes, in particular (Panteleev et al. 1978), this ash is acidic rather than highly mainly, which greatly limits its use in unchanged form as active mineral additive to Portland cements, since it does not initially have With any noticeable hydraulic activity. And besides, it contains certain amount of unburned carbon (coal, semi -lane), content of which for concrete is greatly limited by the requirements of GOST.

Given the foregoing, search for other ways to use ash of the Ekibastuz coals, except to use it as active mineral additive to cements, is a fairly urgent task in the national economy.

One of these paths, in our opinion, is the study of the possibility of obtaining from the Ekibastuz ash of Agloporite—artificial aggregate for light concrete.

Materials and Methods The main purpose of this work was to obtain by firing the TPP-3 ash from burning of Ekibastuz coal with various additives-aggloporite gravel with average density below 1.0 g/cm^3. That is, such density that granules of the aggloporite swim in the water, as well as that they fully satisfy the requirements of the following regulatory documents: GOST 11991-83 "Break stone and agloporite sand. Specifications" and GOST 9757-90 "Gravel, break stone and sand artificial porous.

Specifications

In the work, to reduce temperatures of sintering and melting of the Ekibastuz ash, guided by cord recipes of ordinary window glass, made based on quartz sand.

To do this, two recipes of glass cords were selected for analogues:

1. two -component, in which soda with small amount of Na_2SO_4 (Na silicate glass) is added to quartz sand;
2. three -component, in which in addition to soda and Na_2SO_4 calcium component was also added—$CaCO_3$calcite (Na-Ca silicate glass).

The choice of analogues of glass batch was justified by the following considerations.

Firstly, as described above, the melting temperature of quartz (in the case of building glass-quartz sand) is quite high 1700–1730 °C.

Adding alkaline and alkaline earth components (Na_2CO_3, Na_2SO_4, $CaCO_3$) allow to significantly reduce this temperature to the level 1450 °C.

Secondly, the process of obtaining industrial construction glass is divided into four stages (GOST 8269.0-97. 1998):

1. silicata formation (at temperatures 800–900 °C);
2. glass formation, characterized by significant translation (melting) of previously formed silicates into glass with a large number of bubbles (at a temperature 1100–1200 °C);
3. clarification, in which arrangements added initially and increase in temperature to 1450 °C lead to disappearance of bubbles in the glass;
4. homogenization (or averaging), which occurs at temperatures of about 1400–1500 °C, process of which remains previously undergraved silicates and mixing glass to averaging its composition are melted.

For the purposes of this work, the last two stages were not needed. On the contrary, in our case, it was necessary to stop process at the stage of glass formation at the moment when number of gases in the glass will be maximum.

The choice of not one (two-component Na container) composition, but two compositions, considering the Na-Ca containing composition, was made at the very beginning of study, considering well-known information, which is purely NA-glass, as a rule, more soluble and less chemically rack Compared to Na-Ca glass. And this, in turn, was important to us since one of technical requirements for aggloporite according to regulatory documents was resistance of aggloporite substance to silicate decay.

Thirdly, for our case, positive factor was presence in the Ekibastuz ash as a restorer (coal), which contributed to significant acceleration of decomposition Na_2SO_4 and its interaction with SiO_2, due to which general melting temperature is generally reduced.

In our case, instead of calcified soda Na_2CO_3 the baking soda $NaHCO_3$was used, and Na_2SO_4 was added at the rate of not more than 5% of the mass of soda in accordance with the recommendation (GOST 8269.0-97, 1998).

Research and development of a method for obtaining a filler of light concrete. Two recipes for aluminosilicate glasses were made and tested: Na aluminosilicate and Na-Ca aluminosilicate glass. Aluminosiacs, in our opinion, must be called based on the composition of the Ekibastuz ash, in which aluminosilicates are quantitatively prevailed: sillimanitis and mullit, and quartz has subordinate significance.

Recipes of tested aluminosilicate glasses are presented below:

1. Na-glass:
   1. ash—85,58%;
   2. soda—11,02%;
   3. Na_2SO_4—0,55%;
2. Na-Ca-glass:
   1. ash—93,89%;
   2. soda—5,63%;
   3. Na_2SO_4—0,5%.
   4. Ca_2CO_3—2,77%.

   Recipes were first tested in form of loose batch in tigs at various temperatures in the range from 1200 to 1310 °C and various excerpts. As a result of the most optimal, excerpt was recognized for half an hour at a temperature of 1280 °C which was later used to prepare granules of both types of glasses.Temperratue rise in muffle was 10° per minute (Bek et al. 2022; Nurpeissova et al. 2022).

The main characteristics of glass were determined according to GOST 9758-2012 "Porous inorganic fillers for construction work. Testing methods", and evaluated them according to requirements of GOST 11991-83 "Break stone and agloporite sand. Specifications" and GOST 9757-90 "Gravel, break stone and sand artificial porous. Specifications.

As a result of initial studies, the following was established:

1. The amount of unsecured carbon in the TPP-3 as the results of determination of MLI. according to GOST 21216-2014 amounted to 16,16%. From which it follows that the average amount of non-consumed coal in the Ekibastuz ash TPP-3 can be about 15–17%.
2. On the X-ray diffractometer, the DRON -3 radiogram was obtained by the CHP-3

As a result of the interpretation of this radiograph, the following minerals were installed in the following quantities, in % from the crystalline phase: hematite Fe_2O_3—12,1%, quartz SiO_2—32,4%, sillimanitis Al_2SiO_5—25,9%, mullit $Al_{4,95}Si_{1,05}O_{9,52}$—29,6%.

As can be seen from the above digital values, the secondary minerals formed in the Ekibastuz ash when burning coal to thermal power plants are consistent in quantitatively with the source [9] about number of primary miracles in the Ekibastuz coal.

So, the lion's share (25,9 + 29,6 = 55,5%) is formed by sillimanin and mullit minerals.

3. The TPP-3 ash during withstanding in muffle stove at a temperature of 1310 °C for 1 h showed very weak sintering, small thermal shrinkage and complete absence of at least a slight local melting. In general, material turned out to be sufficiently fragile and loose, relatively easily destroyed with pressure.

This obtaining result is well consistent with the early data of the source (Kuldeyev et al. 2024), in which start temperature of sintering for the Ekibastuz ash was defined as 1300 °C.

From this, we concluded that without additives in this ash alkaline or alkaline earth elements that reduce temperature of sintering and melting, infliputer cannot be obtained at a maximum temperature of 1310 °C.

Photos of granules obtained from Na glass are shown in Fig. 9a, and from Na-Ca-glass in Fig. 9b. As can be seen from figures, the Na glass granules have a light gray color, and the Na-Ca-glass granules are dirty and green.

Figure 10 shows that overwhelming number of granules from the Na glass (Fig. 10a) and from the Na-Ca -glass (Fig. 10b) have density below density of water and swim on it surfaces.

In the photographs on some granules of both sodium and sodium-calcium glass, white sandstones are visible. These are grains of quartz sand of fraction of less than 0,5 mm, which sprinkled with scented substrate in the muffle furnace to limit the "sticking" (baked) of glass granules to the forthcraft (Begentayev et.al 2024; Ashimova et al. 2023).

Next, the main physical and mechanical characteristics of the obtained glasses of two types of aggloporite, which are given in Table 9, were determined.

On the radiographic diffractometer of the DRON-3, radiograms of two obtained glasses shown in Fig. 11 a and b.

When comparing the X-ray of the Na and Na-Ca glass among themselves (Fig. 11 a and b), in the Na- Ca -glass primary quartz disappeared completely (only the molten mulite remained completely), while in the Na-glass, its number, although

Fig. 9 *a*- granules of Agloporite from Na-glass; *b*-granules of the Agloporite from Na-Ca-glass

a b

Fig. 10 Granules of the Agloporite from Na-glass (a) and from Na-Ca-glass (b) swim on the water surface

Table 9 Main physical and mechanical characteristics of glasses of two types of aggloporite

№	Physical and mechanical characteristics, units of measurement	Na-glass	Na-Ca-glass
1.	Average density of gravel, g/cm^3	0,95	0,98
2.	Bulk density of gravel, kg/m^3	505,47	514,92
3.	Brand in bulk density gravel	600	600
4.	Gravel strength when squeezing in the cylinder, MPa	2,99	1,39
5.	Mark Gravel in strength (according to GOST 9757–90)	П250	П100
6.	Mass loss in determining resistance against silicate decay,%	0	0,07
7.	Porosity, %	62,0	60,8
8.	True density, g/cm^3	2,5	2,5

a *b*

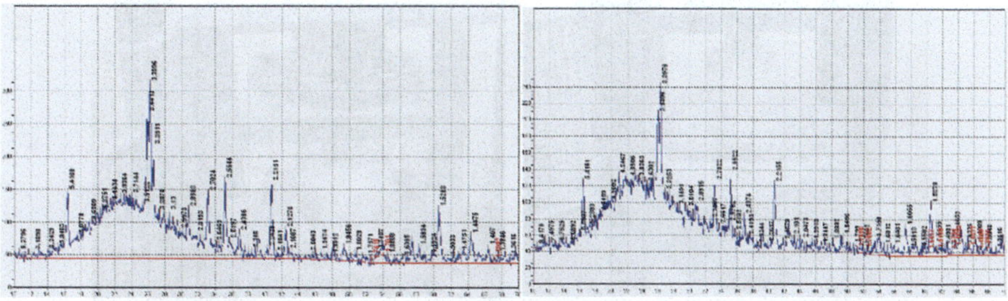

Fig. 11 X-rays Na-glass (A) and X-ray Na-Ca glass (b)

significantly reduced, but entire quartz did not go into glass. It is this factor that can explain the high temperature viscosity of the Na glass compared to Na- Ca -glass.

Determination of bulk density of aggloporite gravel of two types was determined according to paragraph 6 of GOST 9758–2012, and strength when compressed in the cylinder-according to paragraph 25 of this GOST.

In accordance with (GOST 9757-90, 2007) bulk density with values within св. 500 to 600 inclusive—refers to brand in bulk density "600". Therefore, both types of glasses received are attributed to this brand.

According to Table 7 GOST 9757-90 (GOST 9757-90, 2007) agloporite gravel with strength when squeezing in the cylinder within over 1,2 to 1,5 MPa, it belongs to the brand in terms of strength P100, and aggloporite gravel with durability when squeezed in cylinder in limits of 2,5–3,0 MPa-refers to brand in terms of strength P250, which is reflected in the above Table 4.

According to p. 1.3.4 and Table 8 (GOST 9757-90, 2007), brand of aggloporite gravel in bulk density "600" should correspond to brand in terms of strength P100.

Photos obtained on the press of strength data when squeezing in the cylinder, respectively, Na-Ca glass and Na glass are shown in Fig. 12 a and b, respectively.

Determination of resistance of aggloporite gravel against silicate decay was evaluated by alternating steaming and cooling of gravel gravel according to p. 31 of GOST 9758-2012.

According to requirement of p.1.3.7 (GOST 9757-90, 2007), aggloporite gravel should be stable against silicate decay. Mass loss in determining resistance against silicate decay for aggloporite gravel should be no more than 8%.

a) b)

Fig. 12 Indications of strength of aggloporite gravel when squeezing in cylinder (*a*-gravel based on Na-Ca-glass, *b* based on Na glass)

As can be seen from Table 4, agglоporite gravel of two types of glasses, turned out to be very racks against silicate decay and is not subject to dissolution. In this regard, the Na-glass again showed the advantage over the Na-Ca glass.

Comparison of obtained characteristics of two types of aggloporite gravel with the requirements of GOST 11991–83 were not considered here due to fact that the requirements of this standard are lower than requirements of the standard GOST 9757–90).

In addition to above characteristics of obtained glasses, their indicators such as: their middle and true density, as well as their porosity, were also determined. Recall that these indicators GOST 9757–90 are not normalized.

Average density was calculated as follows: average diameters of several granules of one or another type of glass were measured, corresponding volumes of granules were calculated, granules were weighed, and then average density was calculated as arerithmetic of 5 measurements.

As a result, it was found that average density of two glasses are quite close and, accordingly, for the Na and Na-Ca glass amounted to 0,95 and 0,98 g/cm^3.

True density of the two glasses was determined by the accelerated method using the Le Shatelier device according to p. 4.15.2 (GOST 8269.0–97. 1998). It turned out to be for two types of glasses the same and equal to 2,5 g/cm^3.

Porosity of the two glasses was calculated by the values of the average and true density of the corresponding glasses according to clause 12 of GOST 0758–2012.

The actual values of the porosity were close, and accordingly amounted to: 62,0% for Na glass and 60,8% for Na-Ca glass. According to this parameter, the advantage was again behind the Na glass.

Thus, results of the studies showed possibility of using waste as secondary raw ash and slag materials to reduce anthropogenic load on environment.

Two compositions of aggloporites were obtained and tested: Na-glass and Na-Ca glass with a liveliness of 85,6 and 93,9%, respectively.

It is justified to obtain an agporite-a filler of light concrete of a brand 600 in bulk density and brands P100 and P250 in terms of compression in the cylinder that meets the requirements of GOST 9757–90, from the ash of Ekibastuz coal. The technical novelty of obtained aggloporite—aggregate of light concrete is confirmed (Certificate of authorship 2024).

6 Development of Method for Obtaining Non-autoclaved Aerated Concrete

Currently, in Kazakhstan, at a good pace, small-piece block products from gas-and-foam concrete products with an average density of 700, 800 and 900 kg/m^3 are developing. Their indisputable advantages are:

- Simplicity of production technology, which predetermines possibility (using small and medium -sized businesses, including private entrepreneurship) to organize their release as short time, which helps to quickly return capital invested in them and in a timely manner to satisfy the population's need for these products;
- wide base of starting raw materials, including technogenic materials, including fly ash; moreover, production of cellular products based on the latest individuals can be considered the most promising; This contributes to rational distribution of mineral resources, creation of non -waste technology and solving environmental environmental problem.

One of the promising areas of utilization of the fly ash receipt on their basis of cellular concrete, which is due to following circumstances (Autoclaved 1981; Avvakumov and Gusev 2009; Hu 2016):

- a large ash content (its content in composition of cellular concrete can reach 60 … 70% by mass);
- it can be used as a composition of cellular concrete as a siliceous component or as a component in the binder;
- permissibility of content of non-combatant carbon in its composition in a larger quantity (up to 15%), which determines not to enrich it-the last expensive technology itself (for light and heavy concrete is considered better when coal content in the union is less than 3% by mass);
- reduction in cost of production of cellular concrete and contribution of environmental protection;
- obtaining cellular concrete with specified physical and mechanical properties and durability, due to increased activity of fly ash compared to natural silica materials.
- As for cellular concrete itself, now in Kazakhstan, as already said, aerated concrete with a density of 700, 800 and 900 kg/m^3 of non -autoclastic hardening is produced in large numbers, especially for civil construction, although products from cellular concrete mainly They are released in conditions of autoclave hardening (Zbigniew 2014; Vladimirov et al. 2022; Akhmetov et al. 2008).
- From the foregoing follows: availability of raw materials, low fuel intensity, simplicity and organization of production, low cost, relatively high physical and mechanical properties provide ash concrete in a great perspective.

Research Materials Following materials were used as starting materials:

- Portland cement CEM I 42.5 D0 (according GOST 31108-2020), п production of TOO "AlaCem", with the following rounded phase composition, %: alit ($3CaO{\cdot}SiO_2$-C_3S)—58,7; bellite ($2CaO{\cdot}SiO_2$-C_2S)—17,54; felite ($3CaO{\cdot}Al_2O_3$-C_3A) –3,11; celite ($4CaO{\cdot}Al_2O_3{\cdot}$ Fe_2O_3-C_4AF) –15,4; gypsum ($CaSO_4{\cdot}2H_2O$)—5,3;
- fineness of grinding when passing through sieve № 008–97,4%; normal density—28,0%; onset of setting—220 min; compressive strength, MPa, at the age of 7 and 28 days, respectively—29,7 and 45,3 MPa; ultimate compressive strength after heat treatment—29,3 MPa, specific effective activity of natural radionuclides—100,7 Bq/kg;

- fly ash from Ekibastuz SDPP-2 with the following chemical-mineralogical, granulometric and physical characteristics:
- chemical composition, %: SiO_2_56,7; Al_2O_3_28,6; (Fe_2O_3 + FeO)—6,4; CaO—1,1; MgO—0,35; SO_3–1,3; K_2O—0,03; Na_2O—0,52; ппп—3,0;
- phase composition (rounded),%: mullite ($3Al_2O_3 \cdot SiO_2$)—38; α- quartz (SiO_2)-32; силлиманит ($Al_2O_3 \cdot SiO_2$)—12; hematite (Fe_2O_3)- 5; glass phase–10; unburned carbon –3;
- granulometric composition, %: particles up to 0,5 mm—0,14; 0,45 mm– 2,26; 0,25 mm—3,6; 0,1 mm—25,8; 0,09 mm—0,84; 0,08 mm—12,12; 0,06 mm—4,5; 0,05 mm—21,46; 0,045 mm—21,38; 0,04 mm– 7,9; at the same time, unburned carbon is mainly found in large fractions, glass phases are found in small fractions, and minerals are evenly present in all fractions;
- specific effective activity of radionuclides—72 Bq/kg;
- bassanite ($CaSO_4 \cdot 0,5H_2O$)—semi-aqueous gypsum grade not less than G-4 according to GOST 125–2018; its content in the composition of aerated concrete fluctuated within the range of 3…5% by weight;
- aluminum powder according to GOST 5494-2022.

Research Methodology Composition of ash-aerated concretes with average density of 700, 800 and 900 kg/m^3 was developed in accordance with technical requirements of SN 277 80.—Instructions for manufacture of cellular concrete products. Dry mixes with a specific surface area of about 450 m^2/kg were obtained by joint grinding of ash and cement (with the addition of 3…5% gypsum).

Solution mixture was prepared as follows:

- up to 95% of the required amount of water was poured into working capacity of mixer with a speed of 410 rpm, then impeller of mixer was turned on, after which ash, cement and bassanite (semi-aqueous gypsum) were successively fed;
- mixture was stirred for 3–4 min, and then, without stopping, required amount of aqueous suspension of aluminum powder was added, after which the mixture was additionally stirred for 1–2 min.

After mixing, cellular mixture was poured into 10 × 10 × 10 cm molds for 30 s. The molds were filled with the solution in one go. The height of the mold filling with the solution was 9.0…9.5 cm. The molds with the cellular concrete mixture were kept at the pouring posts at room temperature (20…22 °C) during the swelling process. After filling the mold with the aerated concrete mixture and its swelling, the resulting "hump" was trimmed after 2 h. After this, after 3 h, the samples with the mold were placed in a steam chamber and steamed according to the 3 + 8 + 3 h regime (temperature increase to 85 °C + isothermal holding to this temperature + descent).

Physical and mechanical properties of gas ash concrete were determined according to the technical requirements of the relevant standards. Moreover, compressive strength was converted through a coefficient to the strength of reference sample measuring 15 × 15 × 15 cm (GOST 10180-2012).

The work investigated influence of such technological factors as grinding, composition, water temperature and water-solid ratio on the density and strength of aerated ash cement.

Influence of technological factors on density and strength of aerated ash concrete was studied using base composition given in Table 10, from which it is evident that the strength of samples with an average density of 700 kg/m^3 reaches 3,5 MPa, and with average density of 800 and 900 kg/m^3–5,1 and 7,6 MPa, respectively. These indicators confirm that the obtained samples of mechanically activated ash-cement binder, according to technical requirements of GOST 25485-2019, by appearance belong to structural and heat-insulating cellular concrete, have the corresponding grade (class)—M 35 (B 2,5); M 50 (B 3,5) and M 75 (B 5,0).

Shydration of ash-cement mixture and hardening of ash-cement concrete with the acquisition of certain density and strength are significantly influenced by the above factors. We will consider them separately.

Ash used, as mentioned above, is superacidic and therefore low-active. One way to increase its activity is to grind it separately or as part of mixtures. This increases the specific surface area of both ash and cement, which helps to increase their activity (Estemesov & Seitzhanov 2002; Potapov et al. 2016).

Influence of dispersion of ash-cement mixture on strength of gas ash concrete. concrete was determined as follows.

Gas ash concrete with an average density of 700, 800 and 900 kg/m^3 were obtained:

- without grinding the initial ash cement mixture with a specific surface of about 280 m^2/kg;
- with grinding of this mixture to a specific surface of 420…450 m^2/kg.

Results obtained are given in Table 11 and Fig. 11, from which it is evident:

- that strength of samples made from mechanically activated binder is higher by 1.5…3,4 MPa than from non-activated binder;
- that mechanical activation of gold cement binder has a more favorable effect on increasing strength of denser cellular concretes than less dense ones—in the first case, the increase in the strength of samples reaches1.5 MPa, and in the second—3,4 MPa.

Based on the obtained results, dispersion influence of ash cement mixture on strength of gas ash concrete was obtained (Fig. 13).

Table 10 Designed and then refined basic composition of aerated concrete

Given average density kg/m^3	Specific surface area, m^2/kg	Composition, kg/m^2					Compressive strength, MPa
		Ash	Cement	Gypsum	Water	Powder	
700	450	273	363	3,2	368	0,318	3,5
800	430	337	450	3,2	456	0,360	5,1
900	420	351	467	4,1	474	0,4	7,6

Table 11 Effect of dispersion of ash cement mixture on strength of ash aerated concrete

Average density, kg/m^3	State of the mixture	Composition, kg/m^3					Compressive strength, MPa
		Ash	Cement	Gypsum	Water	Powder	
700	without grinding	273	363	3,2	368	0,318	2,0
	with grinding						3,5
800	without grinding	337	450	3,2	456	0,360	3,1
	with grinding						5,1
900	without grinding	351	467	4,1	474	0,4	4,2
	with grinding						7,6

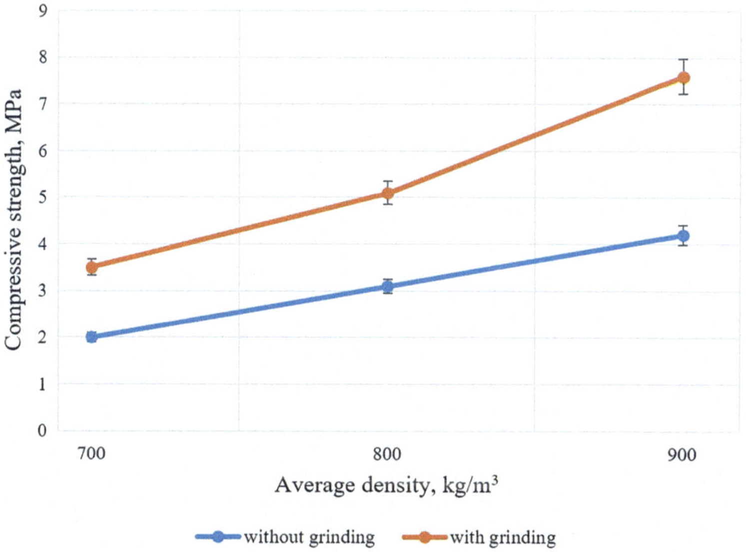

Fig. 13 Effect of dispersion of gold-cement mixture on the strength of gold-cement concrete (Vertical segments represent 95% confidence intervals)

From above it follows that mechanical activation is one of the best ways to improve characteristics of gas ash concrete, since it increases reactivity of mixture components, and it is closely related to well-known patterns of mechanochemical processes of solids, since the following physicochemical and technological factors are implemented (Kuldeyev et al. 2023a, b, c):

- particles are dispersed, as a result of which chemical bonds at crystalline level are broken;
- surface of the particles is amorphized;
- contact between particles increases, which contributes to increase in the density of their packing;
- degree of particles with defects increases.

Table 12 Composition effect of aerated concrete on its strength

Given average density kg/m³	Composition, kg/m³					Average dry density, kg/m³	Compressive strength, MPa
	Ash	Cement	Gypsum	Water	Powder		
700	246	328	2,9	331	0,288	690	3,0
	273	363	3,2	368	0,318	710	3,5
	328	436	3,8	442	0,382	830	4,2
800	303	405	2,9	410	0,32	800	4,2
	337	450	3,2	456	0,360	815	5,1
	404	540	3,8	547	0,432	835	6,0
900	316	420	3,7	427	0,36	890	7,0
	351	467	4,1	474	0,4	920	7,6
	421	560	4,9	569	0,5	940	8,7

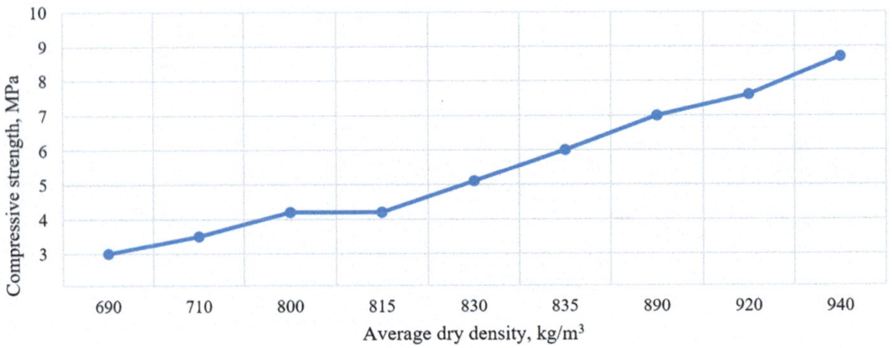

Fig. 14 Effect of composition of aerated concrete on its strength (Vertical segments represent 95% confidence intervals)

Mechanical activation of ash and cement particles in the hardened system interacts with water, forming cementing substances in greater quantities, which is accompanied by increase in the strength of aerated concrete.

Sample Testing Among the main technological factors, composition of cellular concrete is decisive. Since its density, strength and durability mainly depend on the composition (Begentayev et al. 2024).

Table 12 and Fig. 14 show effect of composition on strength of aerated concrete. For each given density (700, 800 and 900 kg/m³), three compositions were selected. Content of components from the base was reduced by 10%, and increased by 20%, respectively. Such combination significantly affects not only strength of the samples, but also affects average density. We will demonstrate this using example of samples with average density of 700 kg/m³:

- when components are reduced by 10% from their content in the base composition, average density decreases from 710 kg/m³ to 690 kg/m³, and the compressive strength from 3,5 MPa to 3,0 MPa;

Table 13 Effect of water-solid ratio (W/S) on density and strength of gas ash concrete

Given average density kg/m^3	Water-solid ratio	Mass mobility according to Sutthardt, cm	Average dry density, kg/m^3	Strength in dry condition, MPa
700	0,52	15	750	3,9
	0,58	17	710	3,5
	0,60	20	690	3,0
800	0,52	14	870	5,9
	0,58	16	815	5,1
	0,60	17	785	4,7
900	0,52	13	980	8,3
	0,58	15	920	7,6
	0,60	16	881	7,0

- when components are increased by 20% from their content in the base composition, average density increases from 710 kg/m^3 to 830 kg/m^3, and the strength from 3,5 MPa to 4,2 MPa.

Analysis of these data shows that the content of components in the composition of aerated concrete should be within the limits of no more than 10% of specified value. Thus, regulation of optimization of compositions of concrete products is one of the main techniques in concrete science. This technique is especially important for cellular concrete, since it is necessary to monitor not only their strength, but also their density.

Influence of Water-Solid Ratio (W/S) Water-solid ratio plays significant role in obtaining aerated concrete with a given density and strength.

The effect of W/S on the density and strength of aerated concrete was studied at its values of 0,52; 0,58 and 0,60. In this case, the following changes occur in the mixtures and aerated concretes (Table 13 and Fig. 15):

- with an increase in the W/S value, as expected, the mobility of the mixture increases, and the average density and strength decrease;
- at W/S = 0,52, samples with a density of 920 kg/m^3 become denser (980 kg/m^3) than samples with a density of 710 kg/m^3 (750 kg/m^3);
- at W/S = 0,60, decrease in density is greater for samples with a density of 920 kg/m^3 than for samples with a density of 710 kg/m^3.

From Table 13 follows Fig. 15—Influence of the water-solid ratio (W/S) on the density and strength of gas concrete.

As can be seen, the water content in the mixture has a very sensitive effect on the density and strength of aerated concrete:

- its decrease increases the strength of aerated concrete (of course, it is favorable), but at the same time its density increases, which is highly undesirable;

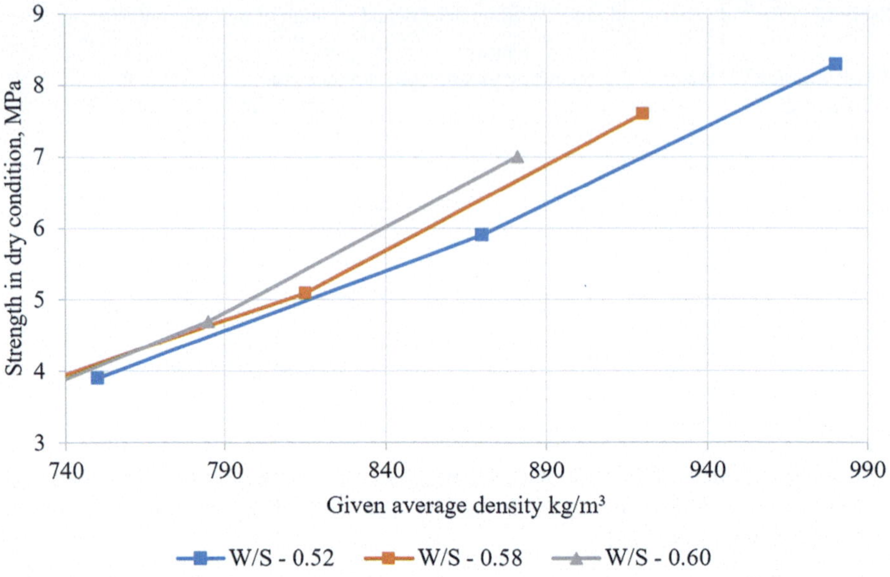

Fig. 15 Effect of water-solid ratio (W/S) on density and strength of gas ash concreted

Table 14 Effect of mixing water temperature on mixture mobility, density and strength of gas ash concrete

Given average density, kg/m³	Water temperature, °C	Mixture mobility, cm	Average density in dry condition, kg/m³	Compressive strength, MPa
700	20	17	710	3,5
	35	15	715	3,8
	50	13	730	4,5
800	20	16	815	5,1
	35	14	825	5,9
	50	11	836	6,8
900	20	15	920	7,6
	35	12	930	8,3
	50	10	939	9,7

- its increase reduces both strength and density of aerated concrete, value of which does not correspond to the technical requirements of GOST 25485-89.

Effect of Water Temperature Table 14 and Fig. 16 show effect of water temperature on mobility, average density, and strength of aerated concrete, from which it can be seen:

- with increase in water temperature from 20 to 50 °C, mixture mobility decreases, and average density and strength of gas ash concrete increases;

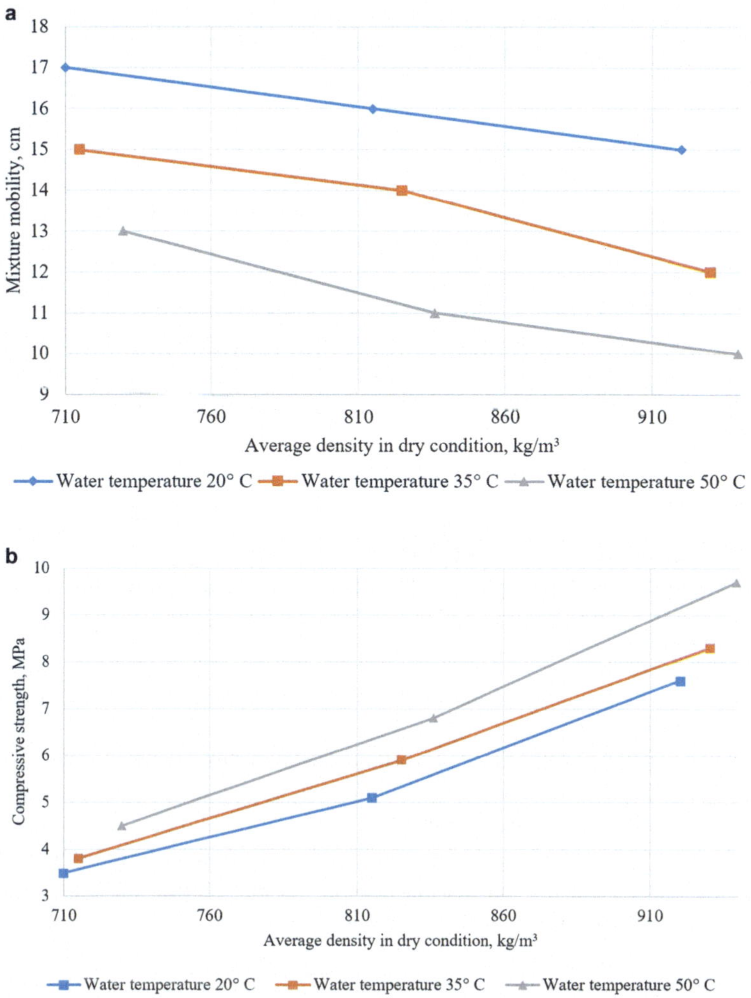

Fig.16 *a*—Effect of mixing water temperature on density and strength of gas ash concrete; *b*—Effect of mixing water temperature on mixture mobility

- with increase in the specified average density of gas ash concrete, effect of elevated temperature has beneficial effect on increasing its strength, which is due to increase in the reactivity of water at elevated temperature, which promotes formation of larger amount of cementing substances as a result of accelerating the interaction between cement and ash particles (Physical and technical properties of raw materials, concrete mixtures and cellular concretes were studied using sand from deposits in Kazakhstan and ash from the Ekibastuz State District Power Plant-2 as fine filler.

It has been established that technological factors of production, such as: dispersion and composition of the mixture, water-solid ratio and temperatures of mixing water have significant impact on density and strength of gas ash concrete, therefore, optimization of these factors allows obtaining gas ash concrete with specified density and strength.

Physical and technical properties of raw materials, concrete mixtures and cellular concretes were studied using sand from deposits of Kazakhstan and ash from Ekibastuz State District Power Plant-2 as fine aggregate. It was found that technological factors of production, such as: dispersion and composition of mixture, water-solid ratio and mixing water temperature have significant impact on density and strength of gas ash concrete, therefore, optimization of these factors allows obtaining aerated concrete with specified density and strength. Technical novelty of obtained gas ash concrete of non-autoclaved hardening is confirmed (Patent No. 8580, 2023).

7 Development of Method for Obtaining Bitumen-Ash-Containing Binder for Road Surfaces

This section discusses use of ash dumps as asphalt pavement in the Almaty region of the Republic of Kazakhstan. Goal is to maintain ecological balance and reduce formation of ash dumps.

In the areas where ash dumps are formed, unfavorable environmental situations are manifested due to large amount of dust formation, as well as the washing out of ash components by wastewater, after which ash components enter soil, because of which mineralization of groundwater increases and environmental degradation is observed. The use of this material will solve environmental problem due to the use of fly ash, as well as reduce evolume of ash dumps, freeing up occupied land for economic needs from ash and.

There are 37 thermal power plants in use in Kazakhstan, which accounts for 72% of the total energy balance. About 72% of electricity in Kazakhstan is generated from coal in the Ekibastuz, Maikubinsky, Turgay and Karaganda basins, 12,3% from hydro resources, 10,6% from gas and 4,9% from oil. Thus, 99,8% of electricity is generated by four main types of power plants, while alternative sources account for less than 0,2%. By burning coal, thermal power plants receive thermal energy and generate electricity. The negative side of this process is the formation of coal combustion by-products—fly ash and slag (Kuldeyev et al. 2023a, b, c; Nurpeissova et al. 2023a, b). About 20 million tons of ash and slag waste are generated in Kazakhstan per year. The main energy companies in Kazakhstan are (Fig. 17):

- Mangistau Atomic Energy Plant—generating company of Aktau, energy supply organization of Mangistau region
- Samruk-Energo—state energy holding

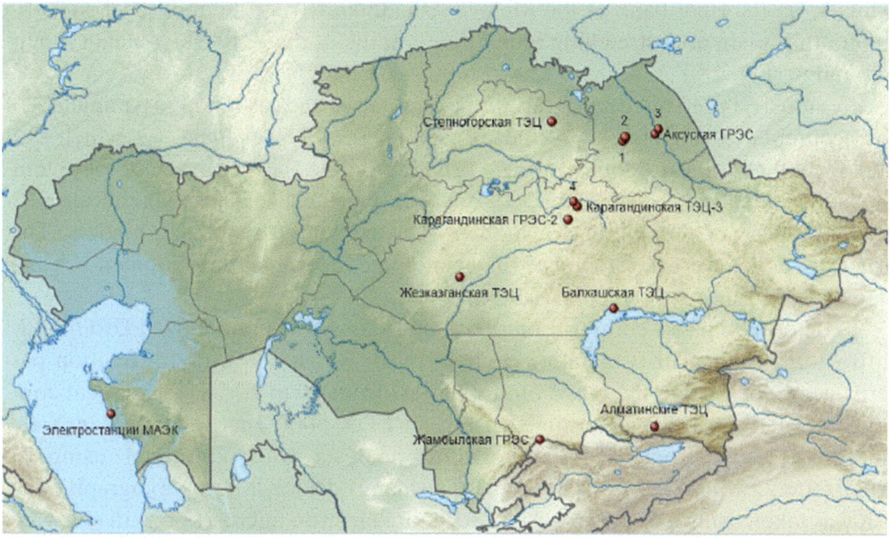

Fig. 17 Map of thermal power plants, state district power plants of Kazakhstan

- KEGOC—national grid operator
- Alatau Zharyk Kompaniyasy—distribution grid company of Almaty
- Almaty EnergoSbyt—energy supply organization of Almaty
- Almaty Electric Stations—generating company of Almaty
- Aktobe TPP—generating company of Aktobe
- AstanaEnergoSbyt—energy supply organization of Astana
- Atyrau Zharyk—distribution grid company of Atyrau
- Uran Energo—grid company

In the areas where ash dumps are formed, unfavorable environmental situations are manifested due to the large amount of dust formation, as well as washing out of ash components by wastewater, after which ash components enter soil, as a result of which mineralization of groundwater increases and environmental degradation is observed. The use of this material will solve environmental problem due to use of fly ash, as well as reduce volume of ash dumps, freeing up occupied land for economic needs from ash and slag mixtures. This paper considers that use of ash as asphalt pavement will not only reduce occupied land, but also improve performance characteristics of road surface.

In the republic, the annual output of ash and ash-slag mixtures from coal combustion is about 19 million tons. There is practically no processing of ash-slag waste on industrial scale. As of 01.01.2022, share of processed, reused, burned ash-slag waste was 0,75%.

One of the largest thermal power plants is JSC Almaty Electric Power Plant—CHPP-3, which provides about 70% of consumers in the Almaty region with electric and thermal energy. The CHPP-3 complex is located 16,5 km from the center of

Almaty, on the right bank of the Malaya Almatinka River. The industrial site is in the foothill plain area stretching along the foothills of the Zailiyskiy Alatau mountain range (Fig. 18.)

Previously, based on the analysis of the qualitative and quantitative parameters of the accumulated ash and slag of the Almaty Thermal Power Plant—3, which uses coal from the Ekibastuz deposit (Kazakhstan), possible areas of their use for obtaining building materials were identified. Ash and slag can also be used in mixtures for forming layers of foundations of highways, as cement-ash and slag binders. This article is devoted to development of technologies for obtaining bitumen-ash and slag binder for use in road construction.

To determine phase composition of ash and slag waste, modernized DRON-3 M diffractometer on CuKα radiation with software was used. X-ray diffraction patterns were obtained in the range of 2θ (angles) from 10 to 70°. The chemical composition of fly ash was determined using an EDX-8000 energy-dispersive X-ray fluorescence spectrometer. Granulometric analysis was performed using an Analizette 22 MicroTec Fritsch GmbH (Germany) device. Microphotograph of fly ash was taken using a Superprobe-733 scanning electron microscope with software.

To determine physical and mechanical properties of the bases using a bitumen ash slag binder, crushed stone gravel sand mixture (CSGSM) is initially mixed with ash and slag at a ratio of 70:30, and then this mixture is heated to 90°. Bitumen is heated separately to 120 °C. Then they are mixed in a ratio of 94:6. Next, samples are prepared from the finished bitumen-mineral mixture by pressing—cylinders measuring 5 × 5 cm. Viscous bitumen BND—90/130 was used to prepare the bitumen ash slag binder (BASB). The strength limit was determined for two types of samples. The first are dry, tested immediately after obtaining samples by pressing, the second are samples that are tested after water saturation.

Fig. 18 Central part of JSC Al ES TPP-3

As a result of the initial studies, the following was established:

1. The amount of unburned carbon in the ash of TPP-3, based on the results of determining the ппп according to GOST 21216-2014, was 16,16%. From which it follows that the average amount of unburned coal in the Ekibastuz ash of TPP-3 may be about 15–17%.
2. X-ray diffractometer DRON-3 was used to obtain an X-ray diffraction pattern of the ash of TPP-3, which is presented in the previous paragraphs in Fig. 6.

The analysis of literary sources on topic of study revealed main areas of application of sludge in production of building products. Mainly as binders, as an active additive in the production of Portland cement, as large and small aggregates, both for concrete, for construction of upper and lower layers of road bases. Developments of building materials available today are not connected by a common direction, are not systematized enough. All this causes an urgent need for targeted comprehensive studies of both the slag itself and materials based on it.

One of the main indicators of raw materials is their granulometric composition. The higher content of micro dispersed particles, the higher the plasticity of the material. Consequently, the raw materials will have high cohesion, which will have positive effect on strength characteristics of finished products, and granulometric composition is also important for determining the adsorption capacity of the material. Results of the particle size distribution are presented in Table 2.

Analyzing the data in this table, we can say the following:

- fly ash is mainly represented by fraction consisting of particles of 100 μm size—25,8%, 80 μm size—12,12%, 50 μm size—21,46%, 45 μm size—21,38%; in total, they reach 80,76%;
- composition of the particle with a size of 500 μm stands apart, which is represented by mullite (28%), α-quartz (54%), carbon (18%) and a small amount of glass phase (occupying a halo area on the X-ray diffraction pattern reaches 6 cm^2; their content from the total amount of fly ash is small—only 0,14%;
- content of unburned carbon in X-ray diffraction patterns is recorded only in the composition of large fractions—from 500 μm to 63 μm; and within these fractions, its content naturally decreases: 18%—7% – from coarse to fine fractions (Aryngazin et al. 2016).

In global construction practice, both unclassified ashes and ashes after their preliminary enrichment or separation into components are widely used. Thus, ashes can be used for construction of road surface bases, in reconstruction of upper worn layers of asphalt as fixative for tar, resin, tar oil for creation of planning embankments, etc. (Sirotok 2010; Osmonova 2015).

Results of the study of physical and mechanical properties of bitumen ash slag binder (BASB) are presented in Table 15.

From Table 15 it is evident that presented test results of samples on bitumen ash and slag binder fully meet the requirements of regulatory documents.

Production of bitumen ash-slag binder and its use in composition of road bases consists of the following main processes:

Table 15 Physical and mechanical characteristics of samples using BASB

Indicators	Unit of measurement	Indicators of physical and mechanical proper	
		Tested samples	ST RK 1225-2013
Compressive strength after 28 days at 20 °C	MPa	2.1	2.0…2.5
Compressive strength after 30 min at 50 °C	MPa	1.9	0.9…1.6
Average density	kg/m^3	2.38	–
Water absorption	%	2.1	1…3

- preparation of bitumen ash and slag binder;
- delivery of binder to the area of construction of motorways;
- preparation of roadbed;
- delivery of sand and gravel or crushed stone-gravel-sand mixture to the place of use;
- displacement of Sand and Gravel Mixture (SGM) or crushed stone-gravel-sand mixture with bitumen ash and slag binder on road;
- formation of the base on the surface of roadbed after the bedding layer or without it;
- compaction of the base using rollers;
- checking density and strength of base for its reliability and stability.

Each of the above-mentioned technological processes is series of individual operations performed in known sequence, with a specific mode on the corresponding equipment.

Technology for producing bitumen ash and slag binder includes the following operations: loading the ash and slag bin; pumping bitumen into the bin and heating it; dosing starting components; preparing bitumen ash and slag binder in a mixer; unloading the finished binder. Figure 19 shows the process flow chart for producing bitumen ash and slag binder.

Homogeneous ash and slag delivered from the warehouse by a cement truck are fed to the bunker (1). Bitumen from the bitumen storage facility is pumped through a pipe to the bitumen melting plant-bunker (2), where bitumen is heated to a temperature of 120–130 °C. From the bunker (1), ash and slag enters the mixer (4) through dispenser (3a), where heated bitumen enters through the dispenser (3b). In the mixer (4), homogeneous mass is obtained from ash and slag and bitumen (in a ratio of 94:6, %) at a temperature of 120–130 °C. The finished БЗШВ is delivered to the site by a dump truck. The type of bitumen may be different depending on the type of base, for example:

Type of bitumen may vary depending on type of base, for example:

- for hot asphalt concrete mixtures, viscous bitumen is used in BZShV (BND—90/130, BND—60/90, BND—40/90), while base is laid at a temperature of 120–130 °C;

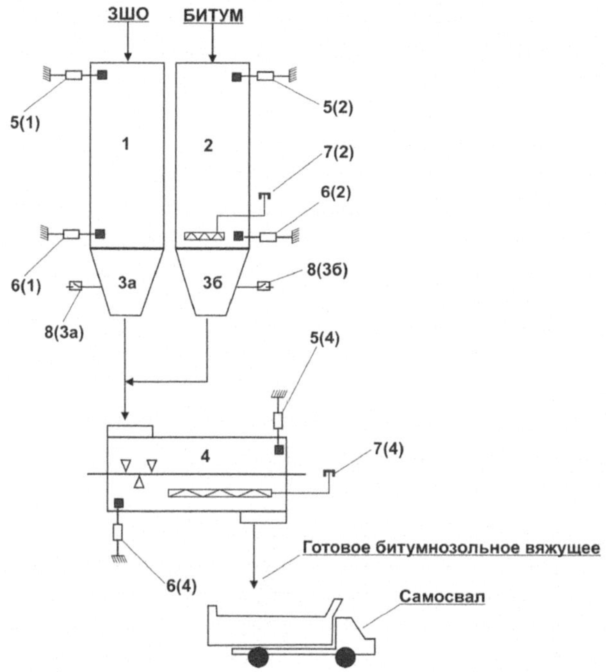

1 – ash and slag waste bin; 2 – heated bitumen bin;
3 – dispensers (3a – ash and slag waste; 3b – bitumen); 4 – heated mixer;
5 – upper level sensors (5(1)-ash and slag waste; 5(2)-bitumen; 5(4)-mixture);
6 – lower level sensors (6(1)-ash and slag waste; 6(2)-bitumen; 6(4)-mixture);
7 – temperature sensors (7(2)-bitumen heating; 7(4)-mixture heating);
8 – dispenser open/close sensors (3a – ash and slag waste: 3b – bitumen)

Fig. 19 Technological scheme for obtaining bitumen ash-slag binder for foundation of highways

- for warm asphalt concrete mixtures, BZShV consists of liquid bitumen (SG—130/120) or low-viscosity bitumen (BND—130/120, BND—200/300), while the base is laid at temperatures of 60–80 °C;
- for cold asphalt concrete mixtures, BZShV is prepared from liquid bitumen (MG—70/130 or SG—70/130) and the base is laid at a temperature corresponding to the ambient temperature (but not lower than +10 °C).

Studies have shown that the introduction of ash and ash-slag mixtures from the combustion of hard and brown coals as a mineral powder for the preparation of asphalt concrete mixtures makes it possible to obtain a material with standard physical and chemical characteristics.

Thus, all conducted studies have shown possibility of using ash and slag materials as secondary raw materials in order to reduce the anthropogenic load on the environment. Based on study of physical and mechanical properties of binder using ash and slag, technology for obtaining a bitumen-ash-slag binder has been developed, the main parameters of which are: type and content of bitumen in the binder, the mixing time in the mixer, the density of bitumen-ash-slag dough (ODM 2013).

Use of ash and slag in road construction makes it possible to completely utilize ash and slag waste from thermal power plants, which helps solve environmental pollution problems. Use of ash and slag helps reduce cost of road surfaces and allows for road bases with significant strength, density and durability. Proposed technical solutions make it possible to create low-waste technologies using ash and slag from thermal power plants, promote the rational use of mineral resources and solve environmental issues.

8 Development of Composition and Method of Preparation of Splitter Block with Addition of Ash

The present invention pertains to realm of construction and manufacturing of building materials, particularly focusing on the formulation of splitter ash bricks, which incorporates ash resulting from the operation of a combined heat and power (CHP) plant. This mineral constituent, a by-product of energy production, is incorporated into the mixture in order to enhance its technical and ecological characteristics.

When utilizing splitterized ash bricks with appropriate reinforcement for constructing walls, a structure akin to stone-monolithic masonry is attained, presenting a number of advantages compared to conventional monolithic structures.

The technical objective is to develop a composition and method for preparing splitter ash brick, incorporating ash derived from a CHP plant, with the aim of improving the thermal insulation, water resistance, and strength properties of the brick, as well as enhancing its environmental sustainability.

The key technical contribution of this invention lies in the development of a composition and process for producing a splitter ash brick that incorporates ash from a CHP facility, resulting in enhanced thermal insulation performance through increased porosity of the material. Furthermore, the incorporation of CHP ash contributes to increased strength of the product by forming a denser and more stable structure. Through the implementation of this formulation, environmental sustainability in production is enhanced by effectively managing waste from CHP facilities.

A prototype for this invention is a raw material mix for manufacturing lightweight ash concrete, comprising, by weight: Portland cement M500 D0 22.50–23.75; river sand 76.25–77.50; sewage sludge ash 5.0–10.0 (by weight of specified cement); and mixing water up to 0.5%. (Patent RU 2738072, IPC C04B 28/04).

However, a limitation of this technical approach is the reduced workability of the concrete mixture, with a Hegermann cone value ranging from 114 to 115 s; the necessity for employing high-quality Portland cement of grade M500 or higher; and an elevated proportion of river sand having a grain size exceeding 1.9.

It is worth noting that there is a dearth of information regarding the utilization of CHP ash for enhancing the physicochemical and mechanical properties of splitter bricks.

The aforementioned technical objective is achieved through a composition for manufacturing splitter ash bricks comprising Portland cement CEM I 42.5, with ash serving as the mineral ingredient, along with fractionated sand, and water, with the following weight percentages of components:

Portland cement: 34
Fractionated sand: 34
Water: 17
Fly ash: 15

In order to obtain comparative data on the strength characteristics of splitter ash blocks under identical conditions, samples in the form of cubes with dimensions of 10 × 10 × 10 cm were prepared from the proposed raw material mixture.

The golden cement paste was prepared manually using a mixing bowl and spatula. The compressive strength testing of the samples was carried out using a hydraulic press at the standard 28-day age of storage in a standard hardening chamber with a temperature of 20 ± 2 °C and relative humidity of 95%.

As can be observed from the proposed methodology, there are several advantages:

- Improvement of the properties of splitter ash blocks through the use of ash.
- Reduction of production costs through the utilization of waste from thermal power plants.
- Environmental sustainability through the incorporation of ash into the construction industry.

Table 16 presents the physicomechanical properties of the proposed formulation.

The method most closely associated with the technical result and adopted as a model for the production of split-ash blocks involves the following stages:

1. The initial step involves the blending of ash, Portland cement, sand, crushed stone, and water.
2. Subsequently, the hydrotreated ash is subjected to enrichment.
3. The resulting mixture is then crushed in a ball mill in conjunction with Portland cement, gypsum, and a plasticizing additive called LSTM-2 until a specific surface area of 4500–5000 cm^2/g is achieved.
4. After that, the resulting mixture is introduced, followed by the addition of 2/3 of the calculated quantity of mixing water, sand, and quicklime.

Table 16 Physical and mechanical properties of the proposed composition

Naming of the indicator	Proposed composition
Average density, kg/m3	730
Compressive strength, MPa	420
Water absorption, no more than %	5

5. The mixture is activated for 1–2 min at a speed of 100–200 revolutions per minute.
6. Limestone crushed stone is then mixed with the remaining mixing water in a concrete mixer for another 1–2 min.
7. The activated mixture is loaded from the dispersant into the concrete mixer.
8. Finally, a homogeneous concrete mixture is obtained through the process of final mixing (RU 2,526,072, IPC C04B 40/00).

The drawbacks of this technique are the low strength and significant plastic deformation of concrete.

The primary goal of the present invention is to enhance the thermal insulation, water resistance, and strength properties of a split ash block, while also improving its environmental durability, accelerating the setting and curing of the ash-concrete mixture, increasing the strength and consistency of ash concrete characteristics, and expanding the range of applications for waste products of anthropogenic origin.

This objective is achieved through a novel approach to the production of split ash blocks, which involves grinding ash in a PM 100 planetary ball mill to achieve better dispersion. CHP ash is then added at a ratio of 15% to ensure optimal results. Next, Portland cement CEM I 42.5 CHP ash and fractionated sand are loaded into a concrete mixer, where they are mixed for 2–3 min until the ash is uniformly distributed throughout the mixture. Following this, water is added gradually and slowly to the mixture until a homogeneous consistency is achieved. The entire process of mixing takes place over a period of 5–7 min.

In essence, this innovation represents a groundbreaking solution in the realm of building materials aimed at enhancing the technical and ecological properties of split ash blocks.

Thus, a Eurasian application has been filed for an invention that represents an innovative solution in the field of construction materials, aimed at improving the technical and environmental characteristics of splitter ash blocks (Eurasian application, KZ2024/0760, 2024).

As a result of using the proposed inventions incorporating fly ash from thermal power plants, the following technical advantages are demonstrated:

- Firstly, these blocks exhibit enhanced strength due to the incorporation of CHP ash into their composition. This significantly increases their resistance to mechanical stress, ensuring the durability and reliability of structures built with them.
- Secondly, the use of CHP ash improves the thermal insulation properties of these splitter ash blocks. This leads to reduced heat loss through walls, thereby enhancing the energy efficiency of buildings.
- Thirdly, the utilization of CHP ash contributes to environmental sustainability by reducing the impact of industrial waste on the environment. This is achieved through the efficient disposal of waste, making the production of these blocks more environmentally friendly.
- Furthermore, the cost-effectiveness of this approach becomes evident as the use of CHP ash lowers the production costs due to its affordability. This makes these blocks a competitive option in the market for building materials.

In conclusion, the integration of CHP ash into splitter ash blocks improves their technical performance, making this invention both promising and relevant in the context of building materials. These split blocks represent a modern and ergonomic solution that combines the strength of concrete with the aesthetic appeal of decorative elements. To date, splitter blocks have been available in a wide variety of colors and textures, showcasing their versatility and performance. Their ease of maintenance, including protection against adverse weather conditions, further enhances their desirability. The high frost resistance of these blocks makes them suitable for use in plinths and fences.

A Eurasian application for the «Composition and Method for the Preparation of Geopolymer Ash Concrete» KZ2024/0761.1 dated 16.09.2024 has also been filed with the Eurasian Patent Office (EAPO).

The invention relates to construction materials, in particular to the composition and preparation method of geopolymer ash concrete, which can be used in the construction of civil, industrial, and hydraulic structures (Eurasian application, KZ2024/0761, 2024).

The technical result of the invention is a composition of a mixture for producing geopolymer ash concrete, comprising fly ash, silica fume, fine aggregate, sodium hydroxide (NaOH), sodium silicate (liquid glass), plasticizer «Neolit 203», and water, with the following ingredient content by weight percentage: fly ash 50–60%, silica fume 5–10%, fine aggregate 20–30%, sodium hydroxide (NaOH) 4–8%, sodium silicate (liquid glass) 5–10%, plasticizer «Neolit 203» 0.5–1%, and water 8–12%.

The preparation method begins with preparing an alkaline solution of sodium hydroxide and sodium silicate. Then, the dry components—fly ash, silica fume, and sand—are mixed until homogeneous, followed by the addition of the alkaline solution, plasticizer «Neolit 203», and the remaining water.

The main advantage of geopolymer ash concrete is its high resistance to aggressive environments such as acids and salts, low shrinkage, and enhanced fire resistance. It serves as an environmentally friendly alternative to traditional cement concrete by reducing carbon dioxide emissions and promoting the recycling of industrial waste.

Additionally, a Eurasian application for the «Composition and Method for the Preparation of Self-Compacting Ash Concrete» KZ2024/0762.1 dated 16.09.2024 has been filed with the Eurasian Patent Office (EAPO).

The invention relates to the construction and development of concrete mixtures, specifically self-compacting ash concrete (SCAC) and its preparation method (Eurasian application, KZ2024/0762, 2024).

The technical result of the invention is the composition of a raw material mixture for producing self-compacting ash concrete, comprising Portland cement, fly ash, coarse and fine aggregates, Superplast-PC hyperplasticizer, and water (water-cement ratio W/C = 0.55), with the following ingredient content, kg/m^3: Portland cement 340–390, fly ash 130–180, sand 700–800, crushed stone 850–1000, Superplast-PC hyperplasticizer 3.5–4.5, and water 190–210.

The preparation method involves preparing the components, sieving the Portland cement and fly ash, dry-mixing them with sand and crushed stone, followed by the addition of water and the hyperplasticizer, and subsequent mixing.

Self-compacting ash concrete exhibits high strength and durability, making it ideal for complex structures, dense reinforcement, and vertical elements. The use of fly ash, a by-product of coal combustion, reduces waste, decreases the carbon footprint, and improves the structure of the cement matrix. Self-compacting ash concrete is resistant to external influences, extending the service life of structures and enhancing construction cost-efficiency.

A Eurasian application for the «Composition and Method for the Preparation of Ash-Fiber Cement Panels for Facades» KZ2024/0763.1 dated 16.09.2024 has also been filed with the Eurasian Patent Office (EAPO).

Ash-fiber cement panels, incorporating fly ash from thermal power plants, are widely used in construction and repair. ***They ensure the durability and high operational efficiency of facade systems*** (Eurasian application, KZ2024/0763, 2024).

The technical result of the invention is the composition of ash-fiber cement panels for facades, comprising Portland cement CEM I 42.5, a pozzolanic additive (fly ash), fractionated sand, plasticizer «Neolit 203», cellulose fiber, and water with the following ingredient content by weight percentage: Portland cement 30–40%, fly ash 20–30%, fractionated sand 20–30%, plasticizer «Neolit 203» 0.5–1.5%, cellulose fiber 5–10%, and water 10–15%.

The preparation method includes thoroughly mixing fly ash, Portland cement, and fractionated sand until a homogeneous mass is obtained, followed by the gradual addition of water containing the dissolved plasticizer «Neolit 203». The mixture is mixed to achieve the required consistency and uniformity. The resulting mixture is compacted in a mold under pressure and hardened through the pressing process.

Such panels feature improved strength characteristics, resistance to external influences, low production costs, and environmental safety. The use of fly ash reduces production costs, minimizes shrinkage deformations, and reduces the risk of cracks.

As a result of using the proposed inventions incorporating fly ash from thermal power plants, the following technical advantages are demonstrated:

- Firstly, these blocks exhibit enhanced strength due to the incorporation of CHP ash into their composition. This significantly increases their resistance to mechanical stress, ensuring the durability and reliability of structures built with them.
- Secondly, the use of CHP ash improves the thermal insulation properties of these splitter ash blocks. This leads to reduced heat loss through walls, thereby enhancing the energy efficiency of buildings.
- Thirdly, the utilization of CHP ash contributes to environmental sustainability by reducing the impact of industrial waste on the environment. This is achieved through the efficient disposal of waste, making the production of these blocks more environmentally friendly.

- Furthermore, the cost-effectiveness of this approach becomes evident as the use of CHP ash lowers the production costs due to its affordability. This makes these blocks a competitive option in the market for building materials.

In conclusion, the integration of CHP ash into splitter ash blocks improves their technical performance, making this invention both promising and relevant in the context of building materials. These split blocks represent a modern and ergonomic solution that combines the strength of concrete with the aesthetic appeal of decorative elements. To date, splitter blocks have been available in a wide variety of colors and textures, showcasing their versatility and performance. Their ease of maintenance, including protection against adverse weather conditions, further enhances their desirability. The high frost resistance of these blocks makes them suitable for use in plinths and fences.

9 Conclusions

1. Thus, ash and slag waste, in terms of its chemical composition and properties, is both a source of soil pollution and a profitable and promising raw material. They can be classified as technogenic mineral raw materials, which, unlike natural ones, accumulate over time and are not depleted, which increases the prospects for their study and involvement in use. The extraction of useful components and complete utilization of ash and slag waste through the use of their beneficial properties and the production of building materials will free up space occupied by dumps, reduce the negative impact on the environment, obtain useful products and reduce the rate of consumption of non-renewable natural resources.
2. Method has been developed for producing an ash-containing binder, including mixing and co-grinding Portland cement and fly ash in a ball mill, characterized in that, in order to increase its strength, it additionally contains the superplasticizer NeoLit-400 with the following content of ingredients, wt. %: Portland cement −70-30, fly ash −30-70, superplasticizer NeoLit-400 (top)—0.3%. Technical novelty of the method for producing ash-containing binder has been confirmed by the Ptant of the Republic of Kazakhstan (Patent No. 8579, 2023).
3. Works of scientists show that ash and slag waste from thermal power plants can replace high-quality raw materials in the production of building ceramics (Kuldeyev et al. 2021; Malchik et al. 2016). The effect of waste modification by grinding on the physical and mechanical properties of the final product is shown using the example of ash from Tomsk State District Power Plant-2. It has been established that the increased content of fine particles of alumina has a positive effect on the water absorption and frost resistance of bricks produced by semi-dry pressing at a pressure of 25 MPa. In this case, the amount of waste ash additive in the charge composition can be increased to 50%. The process of producing ceramic tiles may also involve waste from the fuel and energy complex. It has been shown that by varying the content of the main components, it

is possible to change the color and technical characteristics of the resulting materials.

4. It is justified to obtain agoporite—a lightweight concrete aggregate of grade 600 in terms of bulk density and grades P100 and P250 in terms of compressive strength in a cylinder, meeting the requirements of GOST 9757–90, from the ash of Ekibastuz coal. The technical novelty of the resulting agloporite, a lightweight concrete aggregate, has been confirmed (Copyright Certificate 2024).By studying the physical and technical properties of ash from Ekibastuz State District Power Plant-2, non-autoclave-cured ash and gas concrete was obtained, the technical novelty of which was confirmed by a patent of the Republic of Kazakhstan (Patent No. 8580, 2023).
5. Based on the research conducted, we have developed compositions and methods for obtaining various building materials (splitter block, geopolymer, self-compacting concrete, fiber cement panels for facades) with the addition of ash, for which applications have been filed with the Eurasian Patent Office (EAPO)
6. Rational organization of the waste recycling process in combination with efficient modern equipment makes it possible to obtain products from secondary raw materials at a cost 2–2,5 times lower than for similar products from primary raw materials, with comparable product quality.

References

Abdrakhimov VZ (2018) Reducing environmental damage to ecosystems through the use of inter-shale clay and ash and slag material in the production of lightweight bricks and porous aggregate. Coal 10:77–83. https://doi.org/10.18796/0041-5790-2018-10-77-83

Akhmetov DA, Akhmetov AR, Bisenov KA (2008) Cellular concrete (aerated concrete and foam concrete). Gylym, Almaty. 383 p

Aryngazin KS, Larichkin VV, Aldungarova DK, Tleulesov AK, Beisembaev MK (2016) Use of industrial waste: educational manual, vol 61. Kereku, Pavlodar, p 1188

Ashimova A., Kyldeyev E.Y., Nurpeissova M.B. (2023) Processing of ash and slag waste from thermal power plants with production of building materials The 18th international congress for mine surveying in Xuzhou, China from October 24th till October 29th 2023. pp. 287–295

Avvakumov EG, Gusev AA (2009) Mechanochemical methods of activation in the processing of natural and man-made raw materials. Geo, Novosibirsk. 155 p

Autoclaved cellular concrete/Prepared by the editorial board, chairman G. Bove: Trans. from English – Moscow: Stroyizdat, 1981. 88 p

Begentayev M, Nurpeissova M, Kuldiev E, Nurlybaev R, Bek A (2024) Study of the influence of technological factors on the density and strength of ash-gas concrete. NEWS Nat Acad Sci Republic of Kazakhstan Ser Geol Tech Sci 2(458):45–57. https://doi.org/10.32014/2023.2518-170X.292

Copyright certificate No. 44179 of the Republic of Kazakhstan for a work of science dated 02/02/2024. A method for producing agloporite—a lightweight concrete filler for ash dumps in the city of Almaty. (authors: Nurpeissova M.B., Begentaev M.M., Kuldeev E.I., Nurlybaev R.E., Orynbekov E.S)

Butt YM, Sychev MM, Timashev VV (1980) Chemical technology of binders. Higher School, Moscow. 472 p

Efremenko AS, Khaltaeva EP (2014) Application of ash and slag waste from thermal power plants in the production of high-strength lightweight concrete. Bull ISTU 8(91):86–89

Entin ZB, Nefedova LS, Strzhalkovskaya NV (2012) Ashes from thermal power plants—raw materials for cement and concrete. Cement Appl 2:40–46

Estemesov ZA, Seitzhanov K (2002) Study of the dynamics of accumulation of ash and slag dumps and their impact on the environment. TseLSIM, Almaty. 270 pp

Eurasian application. Authors: Nurpeissova M.B., Begentaev M.M., Kuldeev E.I., Nurlybaev R.E., Orynbekov E.S., Bek A.A. «Development of the Composition and Method for the Preparation of a Splitter Block with the Addition of Ash». KZ2024/0760 dated 16.09.2024. Eurasian Patent Office (EAPO)

Eurasian application. Authors: Nurpeissova M.B., Kuldeev R.E., Nurlybaev R.E., Orynbekov E.S., Bek A.A. «Composition and Method for the Preparation of Geopolymer Ash Concrete» KZ2024/0761.1 dated 16.09.2024. Eurasian Patent Office (EAPO)

Eurasian application. Authors: Nurpeissova M.B., Begentaev M.M., Kuldeev E.I., Zhetpisbaeva A.Zh., Bek A.A. «Composition and Method for the Preparation of Self-Compacting Ash Concrete» KZ2024/0762.1 dated 16.09.2024. Application to the Eurasian Patent Office (EAPO)

Eurasian application. Authors: Nurpeissova M.B., Begentaev M.M., Nurlybaev R.E., Bek A.A. «Composition and Method for the Preparation of Ash-Fiber Cement Panels for Facades» KZ2024/0763.1 dated 16.09.2024. Application to the Eurasian Patent Office (EAPO)

GOST 9757-90 (2007) Artificial porous gravel, crushed stone and sand. Technical conditions. Moscow. Standardinform

GOST 8269.0-97 (1998) Crushed stone and gravel from dense rocks and industrial waste for construction work. Methods of physical and mechanical tests. Moscow. Interstate Scientific and Technical Commission for Standardization, Technical Regulation and Certification in Construction, 47–50

Gritsuk AI, Tumanova SA, Chargazia TZ, Borodai DI, Stukalov AA (2016) Efficiency of using ash and slag in road construction. Econ Construct Urban Manag 141:105–121. https://doi.org/10.1016/j.earscirev.2014.11.016

Shuguang H, Zhou WF (2016) Lightweight concrete: Per. from China. ASV, Moscow. 304 p

Kuldeyev E.I., M.B. Nurpeissova Estemesov Z.A., Lozinsky V.G., Bek A.A., Ashimova A.A. Prospects for the use of ash and slag waste for the production of building materials Almaty: Satbayev University, 2023a., 182 p

Kuldeyev EI, Nurpeissova MB, Estemesov ZA, Bek AA (2023b) Investigation of the density and strength of ash-gas concrete. NEWS Nat Acad Sci Republic of Kazakhstan Ser Geol Tech Sci 5:186–199

Kuldeyev EI, Nurpeissova MB, Yestemesov ZA, Ashimova AA, Barvinov AV (2023c) Obtaining agloporite from ash of Ekibastuz coal selected from ash dump of CRPP-3 of almaty city. NEWS Nat Acad Sci Republic of Kazakhstan Ser Geol Tech Sci 2:142–148. https://doi.org/10.32014/2023.2518-170X.289

Kuldeyev EI, Nurpeissova MB, Lozinsky VG (2024) Highly efficient ash-containing building materials and products. LAP LAMBERT, Deutsschland. 110 p

Madanbekov NZ, Osmonova BZ (2016) Application of mineral powder from fly ash from thermal power plants in Bishkek in asphalt concrete mixtures. Bull KSUSTA 1(51):99–103

Malchik AG, Litovkin SV, Rodionov PV, Kozik VV, Gaydamak MA (2016) Analyzing the technology of using ash and slag waste from thermal power plants in the production of building ceramics. IOP Conf Series Mater Sci Eng 127:012024. https://doi.org/10.1088/1757-899X/127/1/012024

Ministry of Energy of the Republic of Kazakhstan (2022) Official Internet resource [Electronic resource]. http://energo.gov.kz/index.php?id=2960 (access date 09/24/2022)

Nurpeissova M, Nurlybayev R, Orynbekov Y, Iskakov A (2024) Research and use of ash and slag waste for the production of building materials. Min J Kazakhstan 3:29–33

Nurpeissova MB, Yestemesov ZA, Bek AA, Ashimova AA (2023a) The study of ash from coal combustion and their use in the production of construction products. Min Min Dep 4
Nurpeissova M, Estemesov Z, Lozinsky V, Ashimova A (2023b) Industrial waste recycling—one of the key directions of business developmen. NEWS Nat Acad Sci Republic of Kazakhstan Ser Geol Tech Sci 2(458):193–205. https://doi.org/10.32014/2023.2518-170X.309
Osmonova BZ (2015) Increasing the efficiency of using road asphalt concrete by using fly ash as a mineral powder. Innov Sci 12:121–124
Patent RU2 664567C1, IPC S 04V 7/13 and S 04V 7/52. Publ 08/21/2018, Bulletin. No. 24
Petrographic composition of coals from the Ekibastuz coal basin. [electronic resource] (2022). https://fccland.ru/geologiya-mestorozhdeniy/9370-petrograficheskiy-sostav-ugley-ekibastuzskogo-kamennougolnogo-basseyna.html (date of access 12.12.2022)
Patent of the Republic of Kazakhstan 8579 for the utility model "Method for producing ash-containing binder" (authors: Kuldeev E.I., Nurpeissova M.B., Estemesov Z.A., Lozinsky V.G., Bek A.A., Ashimova A.A.) dated October 27, 2023
Patent RU 2738072, IPC C04B 28/04
Patent RU 2526072, IPC C04B 40/00
Patent No. 8580 of the Republic of Kazakhstan for the utility model "Composition and method of producing non-autoclaved ash concrete (authors: Kuldeev E.I., Nurpeissova M.B., Estemesov Z.A., Lozinsky V.G., Bek A.A., Ashimova A. A.) from October 27, 2023
Panteleev VG, Melentyev VA, Dobkin EL (1978) Ash and slag materials and ash dumps, Moscow, pp 46–49. 51 p
Potapov SO, Svirilova MN, Tanutrov IN, Toloknov MN (2016) Physico-chemical properties of fly ash from the combustion of Ekibastuz coals. Butlerov Commun 45(3):36–39
Sirotok VV (2010) Experience and prospects for using ash and slag materials in transport construction. Expansion of the regional raw material base by involving zoo-slag materials into circulation at the thermal power plant of OJSC TGC-11, 37–52
Suleimenov ST (1996) Physico-chemical processes of formation structures in building materials from mineral industrial waste. Manuscript, Moscow. 298 p
Vladimirov DV, Bulyga LL, Nikiforova VG, Akhmetova AA, Menendez Pidal M (2022) Cellular concrete based on ash of Ekibastuz State District Power Plant-2. J Sci Technol Kazakhstan 1:131–141
Yuryev IY, Skripnikova NK, Volokitin OG (2013) Study of the influence of modified ash and slag waste on the properties of fired ceramic products. Vestnik TGASU 4:191–196
Zhalgasuly N, Estemesov ZA, Kogut AV, Tugelbaev AB (2019) Study of ash and slag waste from a thermal power plant of a mining enterprise as a raw material for the production of building materials. Proc Inst Min 89:308–314
Zotkin AG (2014) Concrete with effective additives. Infra-Engineering, Moscow-Vologda. 160 p
Zbigniew G (2014) Fly ash in the composition of cement and concrete: Per. from Polish. Phoenix, St. Petersburg. 189 p

Establishing a Joint Venture for the Production of Construction Materials Utilizing Waste from Mining, Metallurgical Complexes, and Mineral Raw Materials

Abstract This chapter presents a project focused on establishing a joint production facility for the manufacture of modern and in-demand construction materials, including dry building mixtures and splitter blocks. The project involves the construction of production facilities, installation of advanced technological equipment, and implementation of innovative manufacturing technologies to ensure the production of high-quality materials that meet the requirements of the modern construction market. The initiative aims to meet the growing demand for construction materials, enhance raw material efficiency, and strengthen the company's competitive position.

On November 11, 2024, the *TECHNOPARK STROYTECH* Research Laboratory for Construction and Building Materials was established at Satbayev University. The laboratory's primary goal is to provide scientific and technical services to Kazakhstan's construction sector. Its activities contribute to improving construction material quality and fostering innovative approaches in the industry. As a result of the setup, the laboratory was equipped with state-of-the-art instruments enabling comprehensive research on construction materials. Key equipment includes the MCR 302e rheometer (Anton Paar), I-Cal 4000 HPC isothermal calorimeter, JSM-7000 desktop scanning electron microscope (JEOL), and geotechnical modeling software such as PLAXIS and Midas GTS NX. These tools support studies of rheological properties, hydration processes, microstructural analysis, and interaction modeling of structures with soil foundations under seismic conditions. This technological foundation enables the lab to address a wide range of scientific and applied challenges, improving the quality, reliability, and resilience of construction materials and structures.

The chapter also covers the production of dry building mixtures based on diatomite, known for their enhanced performance characteristics. Competitive advantages include high strength and adhesion, optimal workability, cost-efficiency, and resistance to environmental factors. These mixtures exhibit:

Adhesive strength up to 10 MPa and compressive strength up to 300 MPa;
High workability and pot life (up to 90 min);

M. Nurpeissova et al., *Development of Technology for Processing of MMC Waste*, SpringerBriefs in Earth Sciences, https://doi.org/10.1007/978-3-032-12976-5_6

Vapor permeability up to 0.12 mg/m^2·Pa and thermal conductivity ranging from 0.1 to 0.7 W/m·K;

Frost resistance up to 75 cycles and thermal stability from −50 °C to +70 °C.

The product range includes both universal and specialized mixtures (insulating, leveling, and finishing) for interior and exterior applications. The use of diatomite improves the environmental friendliness, energy efficiency, and durability of building coatings.

Additionally, Sect. 5 focuses on the development and production of splitter blocks incorporating ash and slag waste, offering a comprehensive solution for enhancing the strength, energy efficiency, and environmental sustainability of building materials. The blocks demonstrate several advantages: high compressive strength (up to 13.9 MPa at 28 days), optimal density (1594–1598 kg/m^3), effective thermal insulation (λ = 0.621–0.623 W/m·°C), and cost efficiency through reduced production costs and mortar usage. They are also resistant to climatic and mechanical impacts and suitable for various types of construction. The blocks are produced using a vibration forming process based on a specially developed technological scheme.

1 Development of Regulatory and Permitting Documentation for Production Establishment

Production of building materials based on waste from mining and metallurgical complexes and mineral raw materials of Kazakhstan is a significant contribution to the development of sustainable development of construction industry. Activities of the joint venture are aimed at production of splitter blocks and dry building mixtures, which contributes to efficient use of secondary resources and reduction of technogenic impact on environment.

Development of regulatory, permitting and accompanying documentation necessary for production organization of dry building mixtures and construction of a technology park building for the facility, for example, "Operation and maintenance of industrial building", as well as for production of splitter blocks based on ash and slag, was carried out in accordance with the roadmap.

Document regulated the sequence of work stages—from obtaining permits and approvals to implementing technological solutions and launching production. Roadmap use ensured the systematization of work processes, coordination of interaction between participants, control over task completion deadlines and prompt adjustments (SNiP RK. 2006; SNiP RK. 2015; SNiP RK. 2017; SN RK. 2010; SN RK. 2021).

In addition, tool made it possible to increase transparency of project implementation, optimize resource allocation and achieve planned results in a timely manner (Table 1).

Table 1 Roadmap of regulatory, permitting and other documentation for production establishment

№	Description (name)	
1	Development of draft design	
2	Land management project development	
1.1.3	Obtaining technical specifications	Technical specifications for power supply
		Technical specifications for water supply and sanitation (sewage)
		Technical specifications for gas supply
4	Obtaining architectural planning assignment	
5	Design and estimate documentation for construction	
6	Approval of draft design	
7	Green space inspection report	
8	Inventory and forest pathology survey of green spaces	
9	Certificate of presence or absence of green spaces in local government	
10	Protocol for dosimetric monitoring and measurements of radon content and its decay products in the air	
11	Letter on the absence of cattle burial grounds and anthrax burial sites	
12	Construction project examination	

1. *Development of draft design.* Draft design is one of the key stages in the creation of new production complex, forming conceptual vision of future enterprise. Its functional structure, main parameters and technological solutions ensuring efficient operation are determined.

At the preliminary design stage, the layout of production and auxiliary premises is developed with a focus on optimizing technological processes and ensuring comfortable working conditions for personnel. Special attention is given to logistics to enable the efficient movement of raw materials, semi-finished products, and finished goods, minimizing unnecessary time and resource expenditures.

During design process, preliminary selection of equipment is carried out that meets modern requirements for quality, reliability and energy efficiency. Available technological solutions adapted to specifics of production are analyzed. Additionally, specialists in the field of ecology and industrial safety are involved, which ensures that designed facility complies with current regulatory requirements and minimizes possible impact on the environment.

Draft design serves as foundation for detailed design, construction, and commissioning, ensuring efficient use of resources and high level of organization in technological processes.

Basic Parameters of Designed Buildings

Designed buildings are detached two-story structures without a basement, rectangular in plan. Their overall dimensions are, for example:

- 19,8 × 42,0 m;
- 12,0 × 30,0 m.

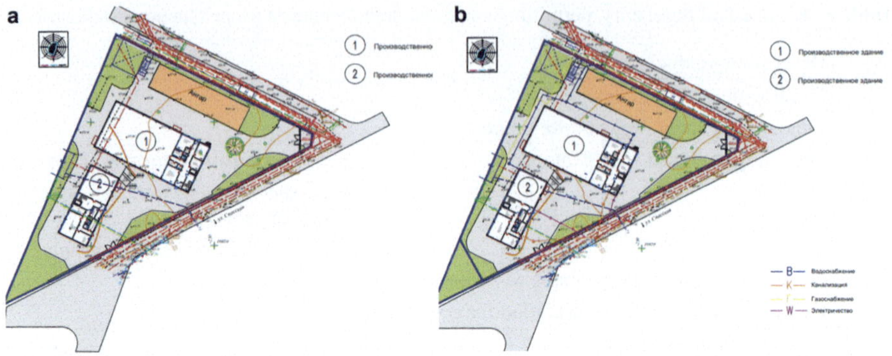

Fig. 1 Operation and maintenance of industrial building. a) General plan Scale 1:1000; b) Utility network plan

General plan with the location of utility networks is shown in Fig. 1.

Design and Technical Characteristics, For Example

1. Load-bearing structure—metal frame with columns and beams;
2. Foundation—monolithic reinforced concrete bases;
3. Roof covering—sandwich panels 150 mm thick;
4. External Walls—Sandwich panels with 100 mm thick basalt mineral wool insulation;
5. Internal walls—cellular concrete blocks 200 mm thick (600 × 200 × 300 mm);
6. Partitions—cellular concrete blocks 100 mm thick (600 × 100 × 300 mm);
7. Stairs—reinforced concrete steps on a metal frame;
8. External metal stairs for fire departments—coating with weather-resistant paints;
9. Porches and ramps—monolithic reinforced concrete;
10. Blind area around the building is 1,0 m wide;
11. Window structures—aluminum profiles with single-chamber glass units;
12. Gates—industrial swing sectional gates of size 3600 × 3000 mm;
13. External doors—metal insulated structures;
14. Internal doors—wooden, blind and metal;
15. Floors—polished concrete with epoxy coating, as well as areas with tiled and linoleum flooring.

Layout of production building and main buildings are shown in Figs. 2 and 3 (Photo 1).

2. *Land management project development.* Development of land management project ensures registration of right to free use of land plot designated for construction and operation of manufacturing enterprise. Selection process considers transport accessibility, proximity to key raw material suppliers, and compliance with environmental and urban development regulations. Comprehensive analysis of territory helps establish optimal conditions for efficient functioning of production infrastructure and logistics processes.

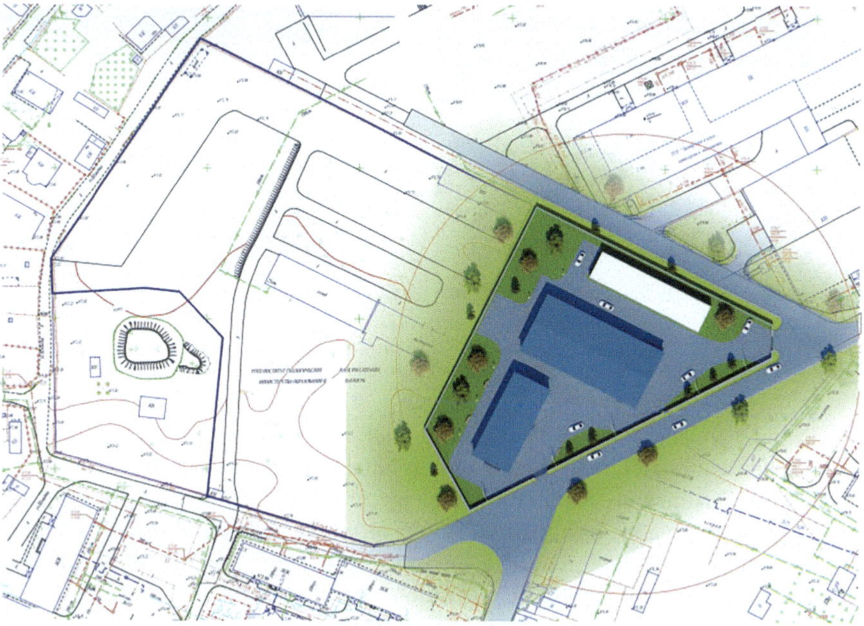

Fig. 2 View of location of production building

Fig. 3 Compensatory tree planting scheme

a) b)

Photo 1 View of main buildings: (**a**) and (**b**) situational view of building

3. *Technical specifications.* To equip a production facility with utility lines, technical specifications (TS) are obtained to regulate the parameters for network connections. Design and coordination work is carried out for water supply, sanitation, power supply, and gasification in accordance with established standards. Connection points are identified, loads are calculated, and measures are developed to ensure the seamless operation of utility systems. Infrastructure is designed with consideration for safety requirements, energy efficiency, and environmental standards, ensuring reliable operation and compliance with the facility's design specifications. Examples include:

Power Supply

- Permitted power—268 kW (380 V);
- Energy supply category—III;
- Permitted power factor—≥ 0,92.

Water supply and sanitation

- Water requirement—1,08 m^3/day;
- Fire extinguishing requirement—30 l/s;
- Guaranteed pressure in the water supply system is 20 m of water column.

Gas supply

- Maximum gas consumption—160 m^3/hour;
- Gas combustion heat—Qp = 8000 Kcal/m^3 (according to GOST 5542-2022).

Obtained technical conditions provided basis for design of utility networks and their subsequent connection.

4. *Receiving architectural planning assignment.* Architectural planning assignment (APA) defines location, parameters and technical conditions of object «Operation and maintenance of industrial building». Document sets out urban planning, environmental and engineering requirements that ensure safety and efficiency of operation.

As part of APA, the following are established: land plot parameters and development standards; requirements for connecting utility lines, including electricity grids, water supply, and transport infrastructure.

5. *Design and estimate documentation for construction.* Design and estimate documentation in the design version is comprehensive document containing calculations, diagrams and drawings necessary for the successful implementation of a construction project. Includes detailed architectural and design solutions, engineering systems diagrams, as well as financial calculations for the estimate part.
   - Architectural and construction part contains information about building layout, design features and appearance. Includes drawings of floor plans, facades and sections.
   - Structural part covers design of load-bearing elements, including foundations, walls, floors and columns, ensuring strength and stability of building.
   - Engineering systems section contains design solutions for water supply, sewerage, heating, ventilation and power supply. Connection points and equipment specifications are indicated, which guarantees the smooth operation of all engineering networks.
   - Estimate section of documentation includes cost calculations for all stages of construction, from preparatory work to final finishing. It accounts for material costs, labor, equipment, and transportation expenses, providing a comprehensive overview of the total construction costs.

Design and estimate documentation is developed in accordance with safety, environmental protection and fire safety requirements. At the stage of the design version, document is approved by the competent authorities.

6. *Approval of draft design* is a key stage of the construction process and is carried out in the Municipal State Institution "Department of Urban Planning and Urbanism of the City of Almaty". Main goal is to ensure that proposed architectural and engineering solutions comply with urban planning regulations, safety requirements, environmental standards and legislative acts. Approval procedure includes checking the design solutions for compliance with established requirements, after which a decision is made on possibility of further design and construction.
7. *Green space inspection report* serves as official document for assessing the condition of trees, shrubs and other plants in an area planned for construction or economic activity. Main objective is to record current condition of plantings and determine need for their preservation, replanting or removal.

   Inspection report ensures compliance with environmental standards during implementation of construction and landscaping projects. Figure 3 shows a diagram of the geolocation of compensatory tree planting. Document was received from the Almaty City Department of Ecology and Environment.

8. *Inventory and forest pathology survey of green spaces* were carried out to determine sanitary condition of trees and shrubs in the area allocated for operation and maintenance of production building. During the work, condition of tree plantations was assessed, potential threats associated with their phytosanitary condition were identified, and recommendations for the care, transplantation or removal of plants were developed. Survey and compilation of taxation description were carried out in accordance with the Rules for the Maintenance and Protection of Green Spaces in Almaty city, approved by Decision No. 211 of the XXXII session of the Almaty City Maslikhat of the VII convocation dated January 17, 2023.
9. *Letter on the absence of cattle burial grounds and anthrax burial sites.* According to inventory and forest pathology survey conducted by a specialized organization (individual entrepreneur or limited liability company), there are no green spaces on the site allocated for construction. This fact is documented and recorded in the established manner.
10. *Protocol for dosimetric monitoring and measurements of radon content and its decay products in the air.* Dosimetric monitoring and measurement of radon and its decay products were carried out to assess the level of radiation exposure to humans and the environment. Radon, as a natural radioactive gas, can accumulate in closed spaces, especially in basements and ground floors of buildings, which creates a potential health hazard.

Measurements help identify radon concentration, determine background radiation levels, and, if necessary, implement measures for their reduction. Regular monitoring of radon content is essential for ensuring the sanitary safety of residential, public, and industrial premise.

Protocol for dosimetric monitoring compiled based on research conducted by the Department of Sanitary and Epidemiological Control of the city of Almaty.

11. *Letter on the absence of cattle burial grounds and anthrax burial sites.* Based on inspection, it was confirmed that there are no cattle burial grounds or anthrax burial sites on the site. This fact is documented and corresponds to current data. Official confirmation was provided by the Almaty City Department of Entrepreneurship and Investments.
12. *Construction project examination.* Expert opinion on construction project contains analysis of compliance of design documentation with current construction regulations, safety requirements and environmental standards. Document may not be partially or fully reproduced, distributed or distributed without written permission of specialized organization that conducted the examination (ExpertCentr LLP or another accredited organization).

2 Establishment of Joint Production Facility for Manufacturing of Innovative and Advanced Building Materials

Creation of joint production facility focuses on manufacturing modern building materials, including splitter blocks and dry building mixes. Project involves construction of production buildings, execution of construction and installation works, and installation of process equipment. Integration of innovative technologies ensures high-quality production, meeting growing demand for building materials and enhancing market competitiveness.

1. *Construction and Installation Works*

Scope of construction and installation works encompasses all stages of production facility construction, from site preparation and foundation laying to the installation of engineering systems and process equipment. All work is performed in accordance with approved design documentation, construction standards, and safety regulations.

Land works—During preparatory stage, construction site is cleared of vegetation, construction waste, and dismantled structures. Geodetic surveys help determine site boundaries, relief levels, and key points for foundation placement. Soil excavation is carried out for foundation, followed by site leveling and removal of excess soil. Final stage includes installing temporary fencing and delivering building materials, ensuring optimal conditions for subsequent construction work (Photo 2).

Photo 2 Land works

Concrete works—After completing land works, a series of measures is undertaken to install concrete structures. Foundation is prepared by compacting soil, laying a sand cushion, and adding a layer of crushed stone to ensure the required bearing capacity. Depending on facility's design features, either a strip or monolithic foundation is used.

Reinforcement meshes and frames are installed, increasing the strength characteristics of concrete and its resistance to loads. Concrete mixture is poured into prepared forms considering design requirements.

Concrete hardening process is carried out in accordance with technological standards, ensuring uniform strength gain. After concrete hydration is complete, foundation is waterproofed, preventing moisture penetration and increasing the durability of the structure Photo 3.

Construction—structural design of building provides for metal frame consisting of columns and beams, as well as monolithic foundations. Roof covering is formed from sandwich panels 150 mm thick. Internal walls are laid out of 200 mm thick cellular concrete blocks, and partitions—from 100 mm thick blocks. Stairs are equipped with reinforced concrete steps on a metal frame. External staircases for fire departments are painted with weather-resistant paints, monolithic reinforced concrete porches and ramps are poured according to design requirements (Photo 4).

2. *Installation of process equipment for dry building mix production.* «White Line» process line for manufacturing dry building mixes using diatomite has been installed. Work includes assembly of main and auxiliary equipment, commissioning activities, and setup of an automated production system. Calibration of batchers, adjustment of mixing units, and testing of transport systems are carried out to ensure seamless operation (Photo 5).

Installed equipment is designed considering physical and chemical characteristics of bulk components, which allows for precise dosing, efficient mixing and stable quality of finished products. Flexibility of process parameters helps to adapt production process to properties of raw materials used, increasing efficiency of processing and system stability. Production capacity of process line is 15,000 tons per year, which allows for uninterrupted supplies of dry building mixtures in required

a) b) c)

Photo 3 Land works. (**a**) foundation preparation; (**b**) concreting; (**c**) concrete elements

a) b)

Photo 4 Metal structure frame: (**a**) frontal view (**b**) profile view

Photo 5 Installation of process equipment for dry building mix production

volumes, meeting needs of large construction companies (RD 34 RK. 2006; Kuldeyev et al. 2024).

Installed equipment «White Line» includes:

1. Bags discharger with a capacity of up to 1,5 m^3–2 pcs;

Photo 6 Installation of technological equipment to produce splitter blocks

2. Screw dispenser SHD-6000/5,5–2 pcs;
3. Dosing hopper BD-1500-2CC with automatic weighing system –1 pc;
4. Screw conveyor SHK-4000 U-4—1 pc;
5. Bucket elevator KT-7000-5,5UD—1 pc;
6. Power frame 3,5 m high (dimensions: 5000 × 4000 × 5000 mm)—1 pc;
7. Mixer for dry ingredients—1 pc;
8. Intermediate bin for storing finished products (volume: not less than 5 m^3, wall thickness: not less than 4 mm)—1 pc;
9. Automatic system for packing in valve bags up to 50 kg—2 pcs with compressor SB3F-300—1 pc;
10. Air cleaning and filtration system with a sleeve self-cleaning filter—1 pc;
11. Belt conveyor for bags with finished products LT-7000-600-4B—1 pc;
12. Automatic control panel for production process PLC-100/2-BL—1 pc.

3. *Installation of technological equipment to produce splitter blocks.* Production line for splitter blocks has a capacity of up to 2500 units per shift, ensuring ability to meet market demand and provide reliable supply of building materials for projects of various scales (Photo 6).

Use of modern technologies and automated processes contributes to high production speed and stable product quality. Optimized work cycles and powerful equipment allow achieving specified production indicators, which is especially important in conditions of increased demand for building materials.

Ash and Slag Grinding Line To prepare raw materials, process complex includes ash and slag grinding line, ensuring its homogeneity and required degree of fractionation. Equipment set includes:

1. Receiving bin—volume not less than 5 m^3;
2. Belt feeder—length 8 m, width 650 mm, equipped with a frequency converter;
3. Vibrating mill—power up to 30 kW, productivity up to 2 t/h;
4. Screw conveyor—length 8 m, diameter 220 mm;
5. Bucket elevator—height 13 m, power 5,5 kW;
6. Silo—volume 40 m^3.

Splitter Block Production Line Production process of splitter blocks is carried out using an automated dosing, mixing and molding system. Equipment includes:

1. Inert material dosing bins with a two-component feeding system;
2. Mixer with a capacity of 750 L;
3. Liquid dispenser with precise parameter adjustment;
4. Automatic cement dosing system with a pneumatic valve;
5. Belt conveyors 500 mm wide, 8 m long, with a power of 4 kW;
6. Vibration press for production of paving slabs, curbs and blocks, equipped with two replaceable matrices;
7. Silo with a capacity of 40 m^3;
8. Screw feeder 6 m long, 159 mm in diameter;
9. Steam chamber for concrete products, area 100 m^2.

Implementation of this technological line significantly enhances production efficiency, reduces manufacturing time, and ensures high-quality finished products.

3 Creation of Laboratory for Certification of Building Materials (TECHNOPARK STROYTECH)

On November 11, 2024, the grand opening of the **TECHNOPARK STROYTECH** construction and building materials research laboratory took place at Satbayev University.

Laboratory is designed to support Kazakhstan's construction sector by conducting comprehensive research and testing of building materials and products using advanced instrumental method and devices (Nurlybayev 2023) (Photos 7, 8, 9 and 10).

TECHNOPARK STROYTECH laboratory is equipped with high-tech instruments for precise research and analysis of various materials. Its advanced

Photo 7 Grand opening of the Research Laboratory of Construction and Building Materials "TECHNOPARK STROYTECH"

Photo 8 Introduction to the laboratory equipment

Photo 9 Opening of mobile workshop for production of splitter blocks in test mode

equipment includes the MCR 302e rheometer by Anton Paar, the I-Cal 4000 HPC isothermal calorimeter, and the JSM-7000 desktop scanning electron microscope by JEOL. Additionally, specialized software such as PLAXIS and Midas GTS NX is utilized for monitoring geotechnical processes, ensuring comprehensive analysis and research capabilities.

Photo 10 Greeting guests at laboratories opening

This equipment enables a wide range of research and development activities, including projects focused on the creation of new building materials and the evaluation of their properties under simulated seismic conditions.

1. The MCR 302e rheometer from Anton Paar is a high-precision device for analyzing the rheological properties of building materials such as self-compacting concrete and mixtures with additives. Rheometer allows for studying viscosity, thixotropy and other characteristics of materials, which is especially important for improving their performance and durability.
2. I-Cal 4000 HPC isothermal calorimeter –device for measuring heat release during hydration of cement compositions. This equipment allows studying processes occurring at stage of cement hardening, which is critical for predicting the strength and stability of concrete structures in the long term.
3. JSM-7000 desktop scanning electron microscope from JEOL—designed for a detailed study of the microstructure of building materials. With its help, it is possible to study surface and internal structure of concrete, composites and other materials at the microscopic level, which helps to identify their defects and improve quality of developments.
4. PLAXIS software—advanced platform for modeling the interaction of building structures with soil foundations. Using this software product, it is possible to carry out calculations and simulations of seismic effects, soil deformations, and assess stability of structures in real operating conditions.

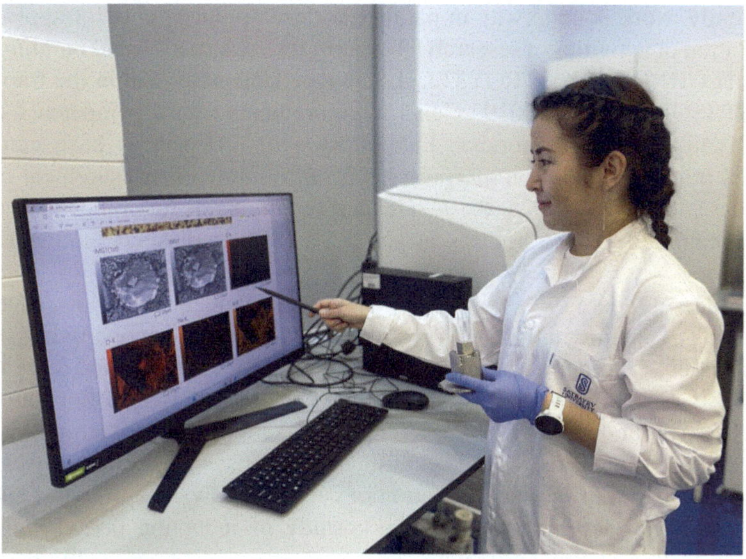

Photo 11 Study of microstructural analysis by young scientists

5. MidasGTS NX—is advanced software designed to perform high-precision calculations and modeling of geotechnical problems. This program is widely used in construction industry to analyze interaction of soils and structures, as well as to assess stability of various geotechnical objects such as foundations, underground structures, tunnels, dams, embankments andslope structures.

Modern equipment allows laboratory to solve a wide range of scientific and technological problems, ensuring high accuracy and reliability of results obtained (Photo 11).

One of the first research project at the laboratory focuses on analyzing the rheological properties of self-compacting concrete (SCC) mixtures for monolithic construction. Successful SCC mix design requires careful consideration of key parameters such as workability, viscosity, and resistance to segregation. The proper selection of components, including microsilica and chemical additives, enhances the water-retaining properties of mixture, minimizing risk of segregation and ensuring stable rheological characteristics. This approach reduces the likelihood of blockages in concrete pumps and disruptions during placement, ultimately improving the efficiency of construction processes.

Laboratory also conducted theoretical and experimental studies in the field of geotechnical seismic isolation, particularly use of vertical barriers made of modified rubber geomaterial (MRG) to protect buildings from seismic impacts. It was found that with optimal barriers, including thickness, height and distance from the building, inertial loads on structures can be reduced by up to 60%. Rubber geomaterials with a 75% increase in rubber lead to a decrease in horizontal accelerations and provide damping of seismic vibrations.

Currently, work is underway in parallel to develop documents for applying for accreditation of activities of research laboratory for construction and building materials of TECHNOPARK STROYTECH Satbayev University within the framework of GOST ISO/IEC 17025-2019 «General requirements for the competence of testing and calibration laboratories». Laboratory staff (10 people) have completed advanced training courses on «Expression of uncertainty of measurement results» and «GOST ISO / IEC 17025-2019—General requirements for competence of testing and calibration laboratories» at the RSE "National Accreditation Center" (Photo 12).

The range of laboratory services: SEM imaging; elemental analysis; EDS mapping; testing of coatings and detection of defects; analysis of nanoparticles; study of materials after heat treatment; measurement of viscosity of liquids and pasty materials; analysis of thixotropy and structure recovery; construction of flow curves for non-Newtonian fluids; long-term tests (creep, relaxation); tests for structural stability and thermal resistance; testing of multicomponent systems (emulsions, suspensions); construction of thermal hydration of cement; analysis of effect of additives (superplasticizers, accelerators, retarders); study of induction period and initial stage of hydration; thermal modeling under various temperature conditions, etc.

Scientific Research Laboratory of Construction and Building Materials TECHNOPARK STROYTECH Satbayev University, as the leading technical university of country, also accepts scientists and young scientists for research internships and exchange of experience. To date, young scientists from the ENU named

Photo 12 Obtaining certificates for group of employees of research laboratory of construction and building materials "TECHNOPARK STROYTECH"

Photo 13 Awarding of certificates of completion of internshipto master's students of Karaganda Technical University (KarTU) named after A. Saginov

after Gumilyov, KarTU named after A. Saginov, Caspian University of Technology and Engineering named after Sh. Yessenov have successfully completed scientific internships in the TECHNOPARK STROYTECH laboratory. In addition, laboratory conducts contract research on topic of «Comprehensive research and assessment of performance properties of modern building materials» and educational practical seminars for manufacturers and employees of research institutes (Nurlybayev et al. 2021) (Photos 13 and 14).

4 Production of Dry Building Mixtures Based on Diatomite

Competitive advantages of dry building mixes:

1. **High strength and adhesion**
   - Cement and gypsum plasters have a bond strength with base from 4(0.4) to 5(0.5) MPa, and tile adhesive—up to 10(1.0) MPa.
   - Compressive strength varies from 10(1.0) to 300(30) MPa, which ensures durability of coatings.
2. **Optimal performance characteristics**
   - High mobility of solutions and ease of application (gypsum plaster—160 ± 5 mm, gypsum self-leveling floor—230 ± 5 mm).

Photo 14 Internship for young scientists of the Caspian University of Technology and Engineering named after Sh. Yesenov

- Long viability of solutions—from 15 to 90 min depending on type of material.

3. **Efficiency and economy**
   - Optimal material consumption due to high plasticity and ease of placement.
   - Possibility of mechanized and manual application to speed up construction work
4. **Vapor permeability and thermal insulation properties**
   - Vapor permeability coefficient up to 0, 12 mg/m^2•Pa, which ensures favorable microclimate in the premises.
   - Thermal conductivity of cement and diatomite plasters within 0,1–0,7 W/m K helps to reduce heat loss.
5. **Frost resistance and resistance to temperature changes**
   - Frost resistance up to 75 cycles, which makes mixtures suitable for use in harsh climatic conditions.
   - Operating temperature range from −50 °C to +70 °C, allowing use of materials in both residential and industrial facilities.
6. **A variety of solutions for different tasks**
   - Mixtures for interior and exterior work, including cement, gypsum, polymer and diatomite compositions.
   - Specialized mixtures: heat-insulating, leveling, finishing and universal compositions for different types of bases (Photos 15 and 16).

Photo 15 Researchers at the construction site of a building materials plant

1. ***Competitive advantages of splitter blocks using ash and slag:***
    - compressive strength after 7 days reaches 7,2–8,3 MPa, and after 28 days—13,4–13,9 MPa, which meets requirements for structural materials and ensures durability of buildings.
    - blocks can withstand significant loads, allowing them to be used for the construction of load-bearing and self-supporting walls.
2. ***Optimum density for strength and thermal conductivity***
   Average density of the material is 1594–1598 kg/m^3, which makes blocks durable and resistant to mechanical impacts, but at the same time relatively easy to transport and lay.
3. ***Effective thermal insulation***
    - thermal conductivity coefficient of 0,621–0,623 W/m °C indicates good thermal insulation properties, allowing to reduce heat loss in buildings and reduce heating costs.

Photo 16 Production of splitter blocks with the addition of ash

- in combination with additional insulating layers, blocks provide high energy efficiency of construction projects.

4. ***Economy and reduction of construction costs***

- use of ash and slag waste reduces cost of block production compared to traditional ceramic and concrete analogues.
- due to precise geometry of blocks, consumption of masonry mortar is minimal, which speeds up construction work and reduces additional costs.

5. ***Environmental safety and sustainable development***

- Recycling of ash and slag waste reduces burden on environment by reducing volume of industrial waste.
- Production of blocks requires fewer natural resources such as sand and cement, making them a more sustainable choice.

6. ***Resistance to external influences***

- blocks are resistant to temperature fluctuations, which makes them suitable for construction in regions with a continental climate.
- high strength of material prevents deformation and cracking during operation.

7. ***Universality of application***

- used for the construction of external and internal walls, partitions, utility and industrial buildings.
- go well with various types of finishing materials, including plaster, facing bricks and ventilated facades.

The blocks are manufactured using vibration molding. Technology for producing building blocks on vibration machines can be presented in the form of the following process flow chart (Fig. 4, 5, 6 and 7) (Nurpeisova et al. 2024a, b).

The uniqueness of the production of splitter blocks is that the organization and launch of the technological line (Fig. 5 and 6).

Houses made of splitter stones do not require cladding. Precise geometric dimensions simplify the masonry process, and the presence of voids allows you to save mortar. Relatively small weight of the blocks does not require a powerful foundation. However, this material is recommended for construction of low-rise buildings. For multi-story construction, it can only be used as cladding (Photo 17).

When constructing buildings, splitter stones can be utilized in various ways:

As a wall structural material for load-bearing and non-load-bearing walls.
As facing products to enhance aesthetic appeal.
As permanent formwork, providing additional strength to structures.

Blocks are laid using either a cement-sand mortar or special masonry glue. Proper alignment is crucial- vertical seams must not overlap. To achieve this, corner and shortened elements are used, eliminating the need to cut full-sized blocks.

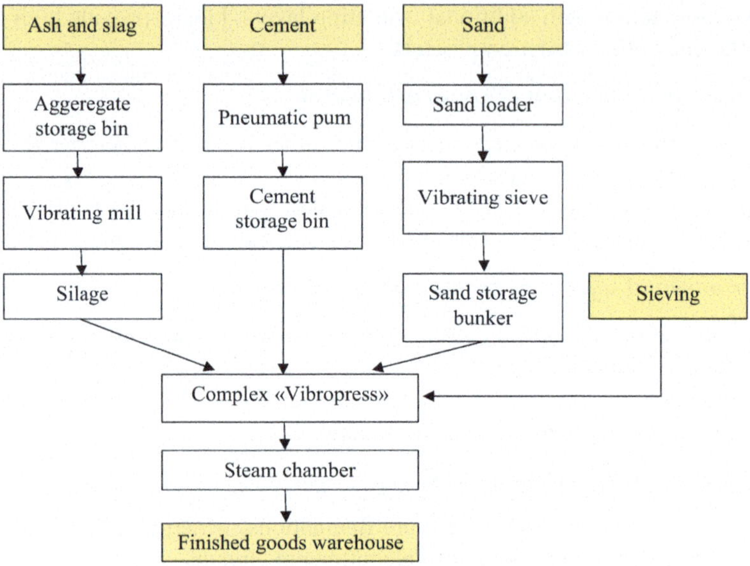

Fig. 4 Technological flow diagram for block production

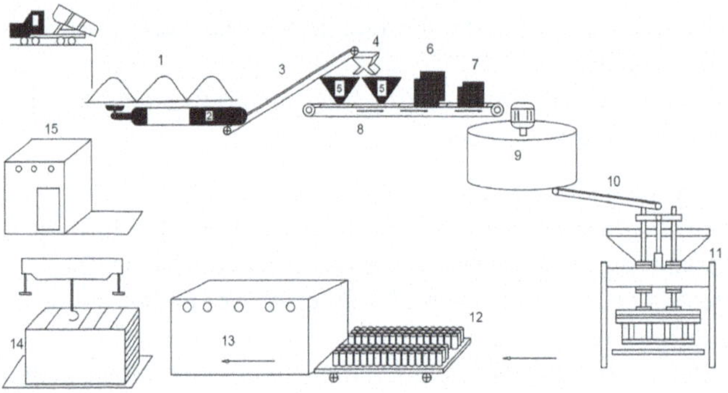

1-bin for receiving ash, slag, sand and screenings; 2-rod mixer for processing ash, slag, sand and screenings mixture; 3-conveyor for delivering processed mixture to the distributor; 4- distributor; 5- bin for storing mixture; 6- bin for storing cement; 7 - bin for additives and pigments; 8 - main conveyor (screw); 9 - concrete mixer; 10- conveyor for delivering finished concrete mix to the press mold; 11 - press mold; 12 - trolley for storing raw products; 13-tunnel steam chamber; 14 - warehouse for finished products; 15 - operator's cabin.

Fig. 5 Technological line of plant for production of building products based on ash and slag cements. 1—bin for receiving ash, slag, sand and screenings; 2—rod mixer for processing ash, slag, sand and screenings mixture; 3—conveyor for delivering processed mixture to the distributor; 4—distributor; 5—bin for storing mixture; 6—bin for storing cement; 7—bin for additives and pigments; 8—main conveyor (screw); 9—concrete mixer; 10—conveyor for delivering finished concrete mix to the press mold; 11—press mold; 12—trolley for storing raw products; 13—tunnel steam chamber; 14—warehouse for finished products; 15—operator's cabin

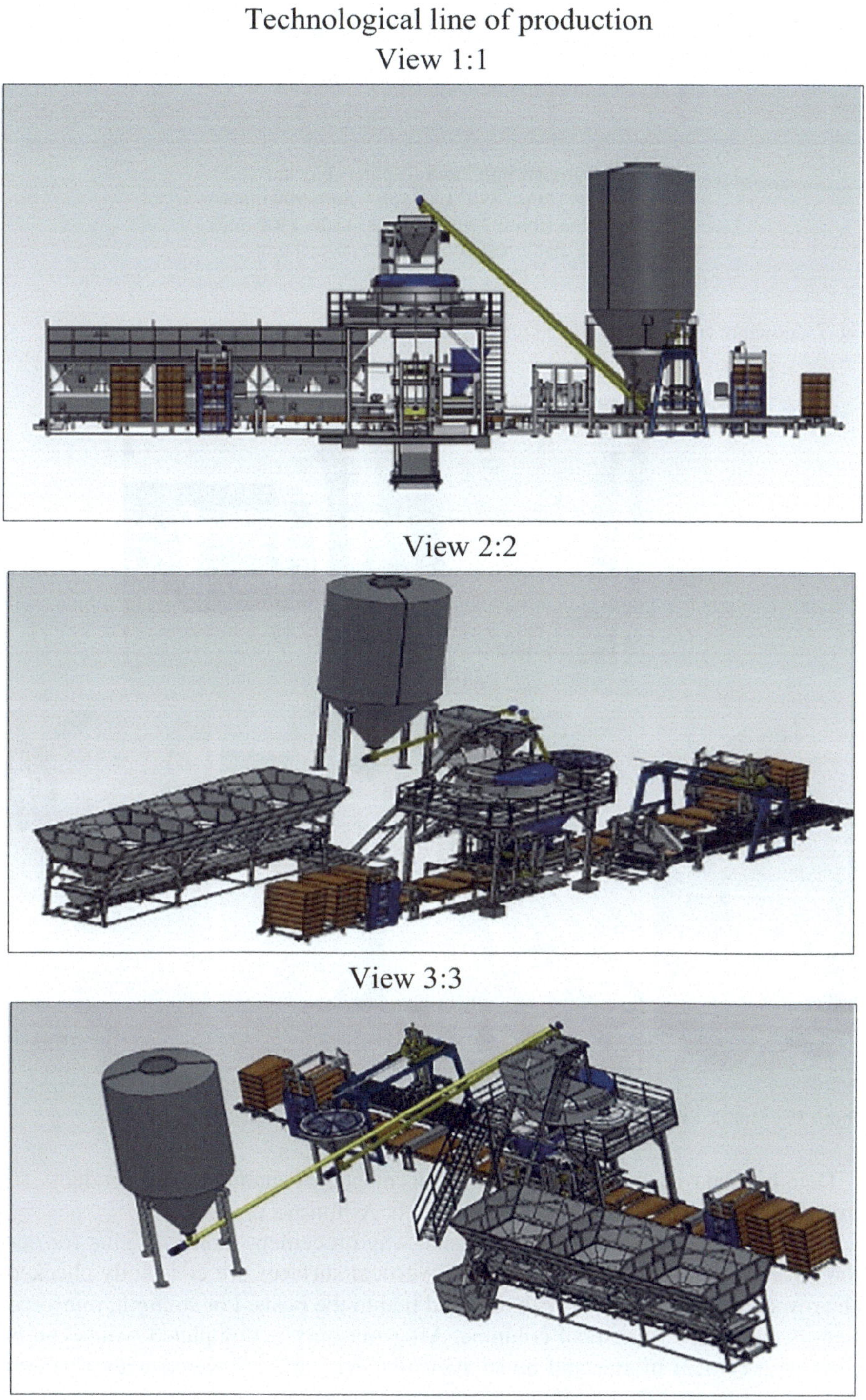

Fig. 6 Technological line of plant for production of building products based on ash and slag cements

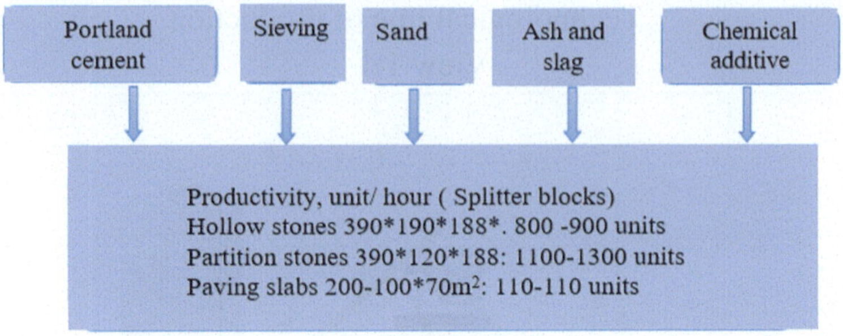

Fig. 7 Structure and productivity of created production

Photo 17 Mobile vibrating machine to produce spitter blocks

Detailed data on composition and performance of manufactured products are provided in Fig. 7 (Nurpeisova et al. 2024a, b; Ashimova et al. 2023).

Masonry process is carried out in usual way on cement mortar or glue for outdoor work. During work, horizontal and vertical surfaces are constantly checked. The rows are also reinforced with nets and tied to the posts. For strength, reinforcement is also made in vertical columns. After masonry is completed, voids can be filled with cement mortar and upper part of fence can be covered with a parapet (Nurpeisova et al. 2024a, b; Begentayev et al 2024).

A fence built from splitter stones looks very similar to a fence made of natural stones. Thanks to a wide range of colors, you can choose the closest to any natural material. And using splitters with a two-sided torn surface, you get a fence with the same finish on the outside and inside.

The solution contains:

- high-quality Portland cement;
- coarse-grained sand or fine-grained crushed stone;
- plasticizer;
- technogenic material, including fly ash;
- pigment for the production of colored products.

For the best results, all components should be mixed in a concrete mixer to achieve a uniform and homogeneous composition. Prepared mixture is then poured into a press form, which must be periodically activated to ensure proper compaction of solution, enhancing strength and durability of final product.

As the blocks are ready, they are placed on a pallet to dry. For better strength gain, it is recommended to water blocks, but only after cement has set. This helps plasticizer in mixture to show its properties to fullest. The blocks gain their design strength faster and more evenly. After 24 h, the finished products can be stacked, and after 48 h they can be transported.

The relief or "split" surface is obtained by using special equipment or by splitting a formed paired product with a special knife. The basic size of a rectangular splitter block is 190 × 190 × 390 mm. For ease of installation, manufacturers produce additional elements of angular shape and other dimensions, as well as partition stones 90 mm thick.

Modern research and technological developments show that man-made waste, including fly ash, can be used as an additive to building materials, including splitter blocks.

The use of technogenic materials, including fly ash, in production of splitter blocks has several advantages, such as reducing the environmental load and increasing the utilization of man-made waste. However, before using these materials in specific projects, thorough study and assessment of their properties, as well as compliance with requirements and standards of construction industry should be carried out (Table 2).

For widespread introduction of this type of raw material, it is necessary to organize production for sorting ash and slag mixtures on hydrovals. The costs of organizing such production will be justified by obtaining a new high-quality fine-grained filler for concrete and products with high consumer properties based on it. Fine-grained concrete based on burnt rocks of mine dumps deserves special attention for the manufacture of paving slabs and paving elements. This is especially relevant for the coal-industrial regions of Kazakhstan, where huge reserves of mine rocks are stored in dumps (Fig. 8).

Thus, processing of technogenic waste can become a promising direction for business development in Kazakhstan to produce competitive building materials and make an important contribution to improving the environmental situation. Use of

Table 2 Technical indicators according to GOST 6133-2019

Name	Unit	Meaning
Strength grade, M25-M300	MPa	2,0-25,0
Frost resistance grades	Cycle	15–200
Water absorption, no more than	%	6
Thermal conductivity	W/(m °C)	0,2-1,0

Fig. 8 Paving slabs

secondary raw materials will reduce need for primary mineral resources, and there will be no need for specialized quarries for development of clays, sands, etc., which disturb the natural landscape.

5 Conclusions on Sixth Chapter

1. Company's activities are aimed at producing splitter blocks and dry building mixtures, which contributes to efficient use of secondary resources and the reduction of technogenic impact on environment. As part of technological process, modern methods of raw material processing are introduced, aimed at creat-

ing building materials with improved performance characteristics. Optimization of formulations will allow achieving high strength, durability and environmental safety of products. This will improve the quality of building products and strengthen their position in the market. Development of these technologies opens new economic prospects, promotes rational use of natural resources and forms innovative approaches in the construction industry.

2. Developed and implemented technological lines to produce dry building mixtures and splitter blocks ensure high dosing accuracy, uniform distribution of components and stable quality of finished products. Equipment adapted to physical and chemical properties of bulk materials optimizes production processes, increasing their efficiency and reliability.
3. Flexibility of process parameters allows adjusting operating modes depending on characteristics of raw materials, which helps to increase productivity, reduce losses and increase range of manufactured products. Production capacity of lines ensures production of significant volumes of dry building mixtures and splitter blocks, which guarantees stable supplies of products for construction market and meets the needs of large enterprises.
4. Automated dosing and mixing systems minimize influence of human factor, ensuring stable quality of manufactured materials. Introduction of modern technologies helps to reduce waste, optimize raw material consumption and increase economic efficiency of production. Thus, integrated approach to equipping production facilities allows improving quality of final product, increasing efficiency of processes and ensuring that manufactured materials meet current requirements of construction industry.

References

Ashimova A, Kyldeyev EY, Nurpeisova MB (2023) Processing of ash and slag waste from thermal power plants with production of building materials. The 18th International Congress for Mine Surveying in Xuzhou, China from October 24th till October 29th 2023, pp 287–295 (In English)

Begentayev M, Nurpeissova M, Kuldiev E, Nurlybayev R, Bek A (2024) Study of the influence of technological factors on the density and strength of ash-gas concrete. News Nat Acad Sci Republic of Kazakhstan Ser Geol Tech Sci 2(458):45. https://doi.org/10.32014/2023.2518-170X.292. (In English)

Kuldeyev EI, Nurlybayev RE, Altayeva ZN, Yestemesova AS, Iskakov AA, Orynbekov ES, Zoldasov AA, Murzagulova AA, Utility model patent RK No. 2024/1032.2 dated 08/26/2024. «Dry construction mix»

Nurlybayev RE, Zhuginisov MT, Zhumadilova ZO, Orynbekov YS, Khamza EE, Sangulova IB (2021) Investigation of the effect of diatomite and bentonite clays on the properties of local loam-based products. Vestnik KazNRTU. KOKHBO 143(4). https://doi.org/10.51301/vest.su.2021.i4.23. (In Russian)

Nurlybayev RE (2023) Scientific principles of obtaining nanodispersed materials based on diatomite for use in the production of heat-insulating materials. Monograph. Under general title "International Education Corporation". 133 p. (In Kazakh)

Nurpeisova M, Nurlybayev R, Orynbekov Y, Iskakov A (2024a) Research and use of ash and slag waste for the production of building materials. Min J Kazakhstan (3):29–33. (In Russian)

Nurpeisova MB, Kuldeyev YI, Bek AA, Ashimova AA (2024b) Rational use of ash and slag waste in the production of advanced building materials. Monograph. Almaty, (In Russian)

RD 34 RK.20.501-04 Rules for technical operation of electrical installations of consumers and Safety rules for the operation of electrical installations of consumers of the Republic of Kazakhstan. Ministry of Energy and Mineral Resources of the Republic of Kazakhstan, dated October 23, 2006, No. 278

SNiP RK 2.03-30-2006 Construction in seismic areas. Department of technical regulation and new technologies is under the jurisdiction of the Committee for Construction and Housing and Public Utilities (HPU) of the Ministry of Industry and Trade of the Republic of Kazakhstan (MIT RK). Effective date: January 7, 2006

SNiP RK 4.02-42-2006 Heating, ventilation and air conditioning. Committee for Construction and Housing and Public Utilities of the Republic of Kazakhstan (with amendments and additions as of October 1, 2015)

SNiP RK 2.04-01-2010 Construction climatology. Department of Technical Regulation and standardization is under the jurisdiction of Committee for Construction and Housing and Public Utilities of the Ministry of Investment and Development of the Republic of Kazakhstan. Effective date: December 20, 2017

SN RK 1.03-15-2009 Instructions for production and acceptance of construction works in cities and industrial enterprises of collector tunnels constructed using shield tunneling method. Department of Scientific and Technical Policy and Standardization of the Agency of the Republic of Kazakhstan for Construction and Housing and Public Utilities (HPU). Effective date: June 1, 2010

SN RK 512-78 Instructions for the design of buildings and premises for electronic computers. Effective date: July 1, 1979. Updated: January 1, 2021

Conclusion

The range of tasks covered in this monograph does not completely cover all the problems of processing technogenic mineral formations (TMF), which are solved using modern methods and technologies

Development of human society is impossible without interaction with the environment, and, therefore, impact on nature without the use of natural resources. However, the urgent need for a fundamental restructuring of modern environmental management technologies has become apparent, driven by the disparate environmental impacts of various industries.

In modern crisis conditions, the decisive factor in increasing business efficiency and meeting growing resource needs is the development and implementation of innovative technologies. This is the most optimal way to develop the economy, since investments aimed at implementing resource-saving measures are most often several times lower than the investments that are necessary to increase the extraction, production and delivery of the required amount of relevant types of raw materials to consumers.

The basis of resource saving is to increase the efficiency of use of material, labor, natural and financial resources, while reducing the total costs of production and sales of a unit of product. Now there is a shortage of natural resources, deterioration of the environmental situation, and resource conservation is becoming a condition for increasing the efficiency of an individual enterprise, associations of enterprises, as well as the construction industry. Accordingly, the widespread introduction of resource-saving technologies in construction will help preserve the natural environment, reduce the cost of construction products, and also improve their quality.

According to the State Cadastre, on the territory of the Republic, about 30 billion tons of industrial waste are stored in dumps, tailing dumps and storage tanks of mining enterprises in Kazakhstan, including: 72%—dump rocks of overburden and substandard ores, 20%—dump tailings, 8%—other waste. With an annual output of

M. Nurpeissova et al., *Development of Technology for Processing of MMC Waste*, SpringerBriefs in Earth Sciences,
https://doi.org/10.1007/978-3-032-12976-5

industrial waste of 1 billion tons. no more than 100 million tons are usefully used. The rest pollutes the environment, gradually accumulating in it.

The integrated use of raw materials and industrial waste from metallurgical, mining and construction enterprises is a pressing problem not only in Kazakhstan, but also in any economically developed state. As practice has shown, waste from these industries is produced in the largest quantities and poses a serious environmental hazard. In these conditions, the problem of environmentally sound use of production waste as secondary raw materials and the development of technologies for popular building materials and products becomes particularly acute.

Development of technologies for the production of building materials based on technogenic waste, contributing to development of industrial and innovative potential of state, careful attitude towards natural resources and the environment, should be considered as the most important scientific and practical task, the solution of which is directly related to environmental safety when disposing of billions of tons of waste in industrial regions. Economic efficiency of waste use is determined by the fact that technogenic raw materials have already been extracted from the subsoil, crushed to a finely dispersed state and placed in dumps.

Processing and recycling of mining and metallurgical waste into construction materials is aimed at solving technological, environmental, social and economic problems in regions with developed mining and metallurgical industries. Rational organization of the waste recycling process in combination with efficient modern equipment makes it possible to obtain products from secondary raw materials at a cost 2–2.5 times lower than for similar products from primary raw materials, with comparable product quality.

Technologies development for production of building materials based on technogenic waste, contributing to development of industrial and innovative potential of state, careful attitude towards natural resources and the environment, should be considered as the most important scientific and practical task, the solution of which is directly related to environmental safety when disposing of billions of tons of waste in industrial regions.

Processing of Technogenic Mineral Formations Will Provide

- reducing costs for searching for new and exploration of exploited deposits;
- preservation of depleting mineral resources in the ground, since the reserves of useful components accumulated in the waste of mining and processing plants are sufficient to satisfy the needs for many decades to come;
- increasing labor productivity due to cost-effective processing of already extracted raw materials, which are essentially finished semi-products;
- improving working conditions, since man-made deposits are located on the surface of the Earth;
- production of cheap building materials (sand, crushed stone, gravel, cement, material for filling the roadway, building dams, dams, etc.), and from slag—(paving stones, tubings, tiles, curb stones, etc.) ;
- it would be even more economically justified if large mining and metallurgical companies consistently created subsidiaries of construction enterprises. Then

our mining and metallurgical complexes can be involved in government projects for the construction of low-rise and affordable housing in the future;

- finally, release of lands it occupies and their reclamation and elimination of sources of environmental pollution, thereby improving the environmental situation around existing enterprises.

Expected effect: Technological effect. As part of the implementation of the state program for industrial and innovative development of the Republic of Kazakhstan in the field of environmental protection, various methods of processing industrial waste through the introduction of waste-free technology are being considered.

Economic effect is the production of additional building materials and products.

Social effect: creation of new jobs in the vicinity of existing mining and metallurgical complexes due to the work of newly created solid waste recycling facilities; improvement of working conditions, since solid waste is located on the surface of the Earth.

Environmental effect: improvement of the environmental situation; reduction of new land areas allocated for dumps; reduction or elimination of costs associated with eliminating the environmental consequences of waste storage.

The manufacturer's authorised representative in the EU is Springer Nature Customer Service Centre GmbH, Europaplatz 3, 69115 Heidelberg, Germany. If you have any concerns regarding our products, please contact ProductSafety@springernature.com

Printed and bound by CPI Group (UK) Ltd, Croydon, CR0 4YY
14/10/2026
02233033-0002